Lecture Notes in Chemistry

Edited by G. Berthier, M. J. S. Dewar, H. Fischer
K. Fukui, H. Hartmann, H. H. Jaffé, J. Jortner
W. Kutzelnigg, K. Ruedenberg, E. Scrocco, W. Zeil

W0246184

8

E. E. Nikitin
L. Zülicke

Selected Topics of the
Theory of
Chemical Elementary Processes

Springer-Verlag
Berlin Heidelberg New York 1978

Authors
Evgueni E. Nikitin
Institute of Chemical Physics
Academy of Sciences of U.S.S.R.
Vorobyevskoye Chaussee 2-b
Moscow V-334/U.S.S.R.

Lutz Zülicke
Central Institute of Physical Chemistry
Academy of Sciences of G.D.R.
Rudower Chaussee 5
DDR-1199 Berlin-Adlershof/G.D.R.

ISBN-13: 978-3-540-08768-7   e-ISBN- 978-3-642-93087-4

DOI: 10.1007/ 978-3-642-93087-4

This work is subject to copyright. All rights are reserved, whether the whole
or part of the material is concerned, specifically those of translation, re-
printing, re-use of illustrations, broadcasting, reproduction by photocopying
machine or similar means, and storage in data banks. Under § 54 of the
German Copyright Law where copies are made for other than private use,
a fee is payable to the publisher, the amount of the fee to be determined by
agreement with the publisher.

© by Springer-Verlag Berlin Heidelberg 1978

Contents

Preface

1.	Introduction	1
2.	Basic Concepts and Phenomenological Description	6
2.1.	Separation of the Center-of-Mass Motion	8
2.2.	Separation of Electronic and Nuclear Motions. Interaction Potentials (Potential-Energy Surfaces)	11
2.2.1.	Heuristic Considerations	11
2.2.2.	Born-Oppenheimer Separation. Adiabatic Approximation	16
2.2.3.	Present State of Potential-Energy-Surface Calculations	23
2.3.	Scattering Channels	26
2.4.	Classification of Elementary Processes. Microscopic Mechanism	27
3.	Dynamics of Atomic and Molecular Collisions: Electronically Adiabatic Processes	32
3.1.	Classical Approach	33
3.1.1.	Some Arguments for the Reliability of the Classical Approach	33
3.1.2.	Atom-Atom Collisions. Elastic Scattering	34
3.1.3.	Quasiclassical Treatment of Elementary Processes in Triatomic Systems: Inelastic and Reactive Scattering	44

3.1.4.   Examples of Results of Trajectory Calculations   59

3.2.   Elements of Quantum-Mechanical Methods   64

3.2.1.   Correspondence of Classical and Quantum-
           Mechanical Theories   64

3.2.2.   Time-Dependent Scattering Theory   71

3.2.3.   Stationary Scattering Theory   77

3.2.3.1.   One-Dimensional Scattering   78

3.2.3.2.   Three-Dimensional Elastic Scattering   83

3.2.3.3.   Rearrangement Scattering (Reactions)   85

3.2.4.   Examples of Quantum-Mechanical Calculations
           of Reactive Elementary Processes   89

4.   Classical-Limit and Semiclassical Approaches
           to the Calculation of Molecular Collisional
           Transition Probabilities   95

4.1.   Classical S-Matrix Method   99

4.2.   The Semiclassical Approach   109

4.3.   Calculation of Transition Probabilities
           Using the Correspondence Principle   116

5.   Theory of Non-Adiabatic Transitions in
           Atomic and Molecular Collision Processes   123

5.1.   Crossing and Pseudo-Crossing of Potential-
           Energy Surfaces   125

5.2.   Non-Adiabatic Coupling and Selection Rules   129

5.3.   The Two-State Problem in Adiabatic and
           Diabatic Representations   135

5.4.   The Landau-Zener Model   145

5.4.1.   Non-Adiabatic Transitions between Electronic
           Terms of Different Symmetry   145

5.4.2.   Non-Adiabatic Transitions between Electronic
          Terms of the Same Symmetry   150
5.4.3.   The Surface-Hopping Trajectory Method   155

Appendix

          Transformation of the Hamiltonian to Center-
          of-Mass and Relative Coordinates   161

Literature   164

Subject Index   171

Preface

Over several decades one realizes a permanently increasing effort to interpret chemical phenomena on the basis of the microscopic behaviour of matter by studying structure and physical properties of atoms and molecules as well as their interactions. In particular, the investigation of elementary chemical processes has become a fascinating field of research being of fundamental importance for chemistry.

Elementary chemical processes connected with energy exchange and redistribution of electronic charge and of atoms depend strongly on the state of aggregation. The theoretical treatment of those processes will be the easier the simpler the conditions are under which they occur. In dilute gases where the interaction time of molecules is much smaller than the time between subsequent collision events, the microscopic dynamics can be treated separately from the statistical problem which leads to a considerable simplification. Correspondingly, the theory of gas-phase processes is comparatively well established at present owing not only to the elaboration of new theoretical methods and computational procedures but also to the development of new, highly sophisticated experimental techniques like molecular-beam and chemiluminescence methods which supplied us with plenty of information about elementary processes between atoms and molecules. It should be pointed out that the success of recent years is to be considered as a result of the joint endeavour of both experiment and theory stimulating each other very effectively. Now the theory of gas-phase elementary processes is reaching a level at which it enables to make not only general statements but also qualitative and quantitative predictions for simple concrete systems. Thus, besides its fundamental importance, research in this field for the future will become practically significant, for example, in some modern branches of chemistry like astrochemistry, air-pollution chemistry, plasmachemistry, and chemical-laser development.

Considering this situation, some knowledge of characteristic features of molecular collision theory is desirable for every theoretical chemist and molecular physicist. The aim of the present course of lectures is to explain the basic concepts of the theory emphasizing the essential suppositions and approximations, mentioning the critical points, illustrating the actual possibilities by examples taken from recent research, and indicating current trends. The presentation will be as obvious as feasible without going into mathematical details.

The whole material is subdivided into five chapters. The first, introductory chapter defines briefly the basic ideas and the terminology of the theory of elementary processes: types of processes, the concept of cross section, and the connection with macroscopic quantities. In the second chapter some fundamental aspects of formulating the theory are discussed, particularly the separation of the center-of-mass motion, the Born-Oppenheimer separation and the concept of potential-energy hypersurfaces, the adiabatic approximation, the notion of scattering channels, and a phenomenological description of some typical mechanisms of processes.

After these more general considerations, in the third chapter the dynamics of simple adiabatic elementary processes (proceeding without change of the electronic state of the total system) is treated in some detail for systems of the type A + BC mainly using the classical trajectory approach; furthermore, elements of the quantum-mechanical description are presented as well.

The classical theory suffers from several shortcomings whereas the application of the quantum-mechanical theory raises so many difficulties that it will be confined to very simple systems and models even in the near future. In the fourth chapter, therefore, we discuss some alternative ap-

proaches, frequently qualified in the literature as classical-limit and semiclassical theories which are able to account
for the substantial quantum effects retaining as much as possible the theoretical simplicity and transparency of the classical approach.

Most important, these methods have the advantage of being,
on principle, also applicable to non-adiabatic processes
which are known to play a dominant role in many chemical reactions. In the fifth chapter this problem is discussed by
means of several semiclassical models.

The reader should not expect to find a full account of modern molecular collision theory; the field is already too
broad and wide-spread. Therefore, concerning more detailed information and some aspects not considered here, it was necessary to refer to the literature; we cited mainly papers of
review character but when concrete systems are discussed,
original papers too.

Several sections of the first chapters are based on lectures given by one of the authors (L.Z.) at Autumn Schools
in Kühlungsborn/G.D.R. 1974 (together with Drs. Ch. Zuhrt
and U. Havemann) and in Jabłonna/Poland 1975. The greater
part of the present text has been prepared during a period
of common work of the authors in summer, 1976; one of the
authors (L.Z.) acknowledges gratefully the opportunity of a
stay at the Institute of Chemical Physics of the Academy of
Sciences of U.S.S.R. and thanks the colleagues in Moscow for
their support and hospitality.

Both authors are indebted to Dr. Ch. Zuhrt/Berlin for
valuable assistance and proof-reading, to Mrs. I. Krüger/
Berlin for typing carefully the manuscript and to Mrs. U.
Schulz/Berlin for drawing the figures; moreover, the authors
are highly obliged to the members of their groups in Moscow
and Berlin for fruitful discussions.

Moscow, August 1976   E. E. Nikitin
                                                   L. Zülicke

1. Introduction

Macroscopic elementary reactions occur as results of statistical sums and averages over various microscopic elementary processes between the individual atoms and molecules of reacting substances. Here we confine ourselves to the case of two s u b s t a n c e s P and Q (reactants) undergoing an elementary gas-phase reaction of the type

$$P + Q \longrightarrow X + Y + \ldots \qquad (1)$$

leading to the substances X , Y , ... (products).

Assuming that, analogous to the macroscopic stoichiometric relation (1), product molecules X , Y , ... in each case are formed by collisions of two reactant p a r t i c l e s   (molecules or atoms), the reaction (1) is called bimolecular.

A stable free molecule is characterized essentially by the quantum states of electronic motion, molecular vibration and molecular rotation[1]. We denote such internal state of a molecule collectively by a letter $i \equiv (n, v, J)$ where n, v and J are the electronic, vibrational and

---

[1] We suppose here separability of these forms of motion and disregard nuclear spin.

rotational quantum numbers, respectively[1]. Depending on the
type of interaction forces between the partners P (in
state i) and Q (in state j), on the relative translational energy, the relative spatial orientation etc., a
collision process can proceed in different ways. In case
that neither the composition nor the internal state of the
particles but only the direction of their relative motion
change, we have an elastic process. If the particles P
and Q retain their atomic composition but change the
internal states, the process is called inelastic. If, by
rearrangement of constituents (atoms, atom groups) of P
and Q , chemically new entities X , Y , ... occur, the
process is a reactive (or dissociative) one:

$$P(i) + Q(j) \quad\longrightarrow\quad \begin{matrix} P(i) + Q(j) & \text{elastic} \\ P(i') + Q(j') & \text{inelastic} \\ X(l) + Y(m) +... & \text{reactive.} \end{matrix} \qquad (2)$$

Besides, there exist other types of processes (ionization,
charge transfer etc.) which will not be considered here.

Given definite collision energy and internal states of
the reactant molecules, usually several of such elementary
processes are possible. A measure of the probability to
observe one of them is the so-called cross section. To
explain this conception we consider an idealized experiment
(fig. 1.-1). We imagine two beams of particles P and Q

---

[1] The symbols n , ν and J generally stand each for
several quantum numbers: n includes spatial symmetry,
spin etc., ν represents the set of normal vibrations.

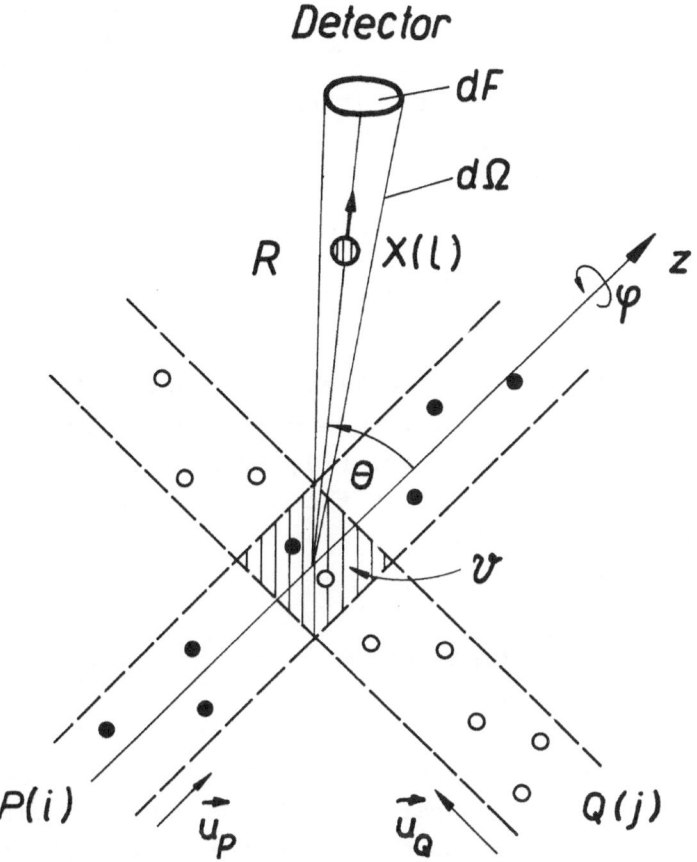

Fig. 1.-1

striking one another; P and Q are assumed to be uniformly in definite internal states, i and j, and to move with uniform directions and velocities, u_P and u_Q, respectively.

By means of a gliding detector at distance R the kind of products occurring as well as their internal states, velocities and directions of motion can be observed. The number $d\mathcal{N}$ of product particles, e. g. $X(l)$, detected per unit time is proportional to the particle density ϱ_P in the beam of particles P, to the particle density ϱ_Q [1], to the relative velocity $u = |\vec{u}_P - \vec{u}_Q|$ of the two particle beams, to the volume V of the interaction zone and to the solid-angle element $d\Omega = dF/R^2$ determined by the detector aperture:

$$d\mathcal{N} \sim \varrho_P \, \varrho_Q \, u \, V \, d\Omega \, . \tag{3}$$

The proportionality factor is called differential cross section of the process considered - creation of product particles $X(l)$ according to eq. (2); it is a function of the relative velocity u and of the angular position ϑ, φ of the detector,

$$\sigma_{P(i), Q(j) \to X(l), Y(m)}(u; \vartheta, \varphi) \equiv \frac{d\mathcal{N}}{\varrho_P \, \varrho_Q \, u \, V \, d\Omega} \, , \tag{4}$$

having the dimension of an area. By summation of all product particles detected in arbitrary directions one gets the total cross section of the process under investigation:

$$\sigma_{P(i), Q(j) \to X(l), Y(m)}^{total}(u)$$
$$\equiv \iint \sigma_{P(i), Q(j) \to X(l), Y(m)}(u; \vartheta, \varphi) \, d\Omega \, . \tag{5}$$

---

[1] The particle densities should be sufficiently low to exclude interactions of particles within the beams.

The aim of the theory of elementary processes is to calculate and to interprete cross sections on the basis of the
interaction forces between the constituent atoms of a
reacting molecular system.

Using these cross sections, one can determine the rate
constant of the macroscopic elementary reaction (1) by
appropriate summation and averaging procedures. In the most
simple case of a complete equilibrium distribution of velocities and internal states of the reactant molecules one
arrives at an integral expression,

$$k(T) = \frac{1}{kT}\left(\frac{8}{\pi kT\mu}\right)^{\frac{1}{2}} \int_0^\infty \overline{\sigma}_{react}(E_{tr}) E_{tr}\, e^{-\frac{E_{tr}}{kT}} dE_{tr} \quad (6)$$

(k being the Boltzmann constant), for the rate coefficient,
where $\overline{\sigma}_{react}$ as a function of the relative collision energy
$E_{tr} = \mu u^2/2$ (μ = reduced mass of the particle pair P-Q)
denotes the total reaction cross section (5) summed over
the product internal states l, m and averaged over the
reactant internal states i and j ,

$$\overline{\sigma}_{react}(E_{tr}) \equiv \sum_l \sum_m \sum_i \sum_j w_i\, w_j\, \sigma^{total}_{P(i),Q(j)\to X(l),Y(m)} \quad (6a)$$

using Boltzmann weighting factors w_i and w_j , respectively (compare e. g. /1/).

2. Basic Concepts and Phenomenological Description

As a first step in the theoretical treatment of elementary processes between atoms and molecules, we are going to discuss some principal aspects of molecular collisions. In the most general description one has to consider a system of two colliding atoms or molecules P and Q , as an aggregate of nuclei and electrons interacting by Coulomb and spin-dependent forces; henceforth, however, the latter will be disregarded and we confine ourselves to non-relativistic theory. All particles are treated as mass points; the system is supposed to be isolated (no external forces).

We introduce an arbitrary, space-fixed coordinate system ("laboratory system") in which the positions of the N nuclei $A, B, C, \ldots$ are given by vectors $\{\vec{R}_k\} \equiv R$ and the positions of the N_e electrons by $\{\vec{\xi}_\varkappa\} \equiv \xi$. The Hamiltonian of this system is composed of five terms,

$$H(R,\xi) = T(R) + T(\xi) + U(R) + U(\xi,R) + U(\xi), \quad (I)$$

representing the kinetic energy of nuclei and electrons as well as the nuclear, nuclear-electronic and electronic interactions, respectively (in non-relativistic treatment):

<table>
<tr><td>classical</td><td>quantum</td></tr>
</table>

$$T(R) = \sum_{k=1}^{N} (\vec{p}_k^{\,2}/2m_k) , \qquad \hat{T}(R) = -(\hbar^2/2)\sum_{k=1}^{N} (\vec{\nabla}_k^{\,2}/m_k), \quad (Ia)$$

$$T(\xi) = \sum_{\varkappa=1}^{N_e} (\vec{p}_\varkappa^{\,2}/2m_e) , \qquad \hat{T}(\xi) = -(\hbar^2/2m_e)\sum_{\varkappa=1}^{N_e} \vec{\nabla}_\varkappa^{\,2} , \quad (Ib)$$

$$U(R) \;=\; \mathring{e}^2 \sum_{k<l=1}^{N-1}\sum^{N} Z_k Z_l / |\vec{R}_k - \vec{R}_l| , \qquad \text{(Ic)}$$

$$U(\xi, R) = -\mathring{e}^2 \sum_{k=1}^{N}\sum_{\varkappa=1}^{N_e} Z_k / |\vec{\xi}_\varkappa - \vec{R}_k| , \qquad \text{(Id)}$$

$$U(\xi) \;=\; \mathring{e}^2 \sum_{\varkappa<\lambda=1}^{N_e-1}\sum^{N_e} 1 / |\vec{\xi}_\varkappa - \vec{\xi}_\lambda| , \qquad \text{(Ie)}$$

where m_k, Z_k and $\vec{P}_k$ denote the mass, atomic number and canonical momentum of the k-th nucleus, m_e and $\vec{P}_\varkappa$ the mass and canonical momentum of the $\varkappa$-th electron; $\mathring{e}$ is the elementary charge. The Laplacians $\vec{\nabla}_k^2$ and $\vec{\nabla}_\varkappa^2$ act on the coordinates of the k-th nucleus and $\varkappa$-th electron, respectively.

The fact that the electronic and nuclear masses are different by at least three orders of magnitude, $(m_e/m_k) < 10^{-3}$, leads to some important simplifications:

(1) The centers of mass of the system or its constituents practically are determined by the nuclei only.

(2) For not too high collision energies (in particular, from thermal energies of the order 10^{-2} eV up to several eV) the electrons move much faster than the nuclei. As a consequence, the nuclei can be treated frequently as playing the role of force centers which change their positions very slowly (adiabatically) so that the electron cloud is able to adjust its state instantaneously to the nuclear framework. On the other hand, the nuclear motion will not be influenced by the momentary spatial arrangement of the single electrons but only by the mean force field (averaged over many periods of the motion) of the whole electron cloud. Under such conditions an approxi-

mate separate treatment of electronic and nuclear motions
is possible (Born-Oppenheimer separation /2/) which will
be discussed in more detail in section 2.2.2.

2.1. Separation of the Center-of-Mass Motion

The substantial aspects of a molecular collision process -
the resulting products, their internal states, the total
cross sections - are determined solely by the motion of the
constituent particles relative to each other but do not depend
on a translation of the interacting system $\{P,Q\}$ as a whole.
Therefore, the first step in the theoretical treatment should
be the elimination of such physically irrelevant overall
translation.

To this end we introduce the center of mass (CM) of the
system $\{P,Q\} \equiv \{A, B, C, ...\}$ defined by

$$\vec{S} = \left(\sum_{k=1}^{N} m_k \vec{R}_k + \sum_{\varkappa=1}^{N_e} m_e \vec{\xi}_\varkappa \right) \Big/ M , \tag{1}$$

if we denote the mass of the nuclei of atoms $A, B, C, ...$ by
$m_1, m_2, m_3, ...$ respectively, and the total mass ,
$\sum_{k=1}^{N} m_k + N_e m_e$, by M. Instead of the laboratory coordinates $\vec{R}_k, \vec{\xi}_\varkappa$ ("L system") we can change to new coordinates, three of them representing the center-of-mass posi-
tion $\vec{S}$. For the choice of the remaining $3N + 3N_e - 3$
coordinates there are several possibilities /3/ to be used
according to convenience. One way is to relate the positions
of the nuclei to one (arbitrarily) selected nucleus, say
the N-th, and the electron coordinates to the center of
mass:

$$\vec{R}_k' = \vec{R}_k - \vec{R}_N \quad (k = 1, 2, \ldots, N-1)^{1)}, \qquad (2a)$$

$$\vec{\xi}_\varkappa' = \vec{\xi}_\varkappa - \vec{S} \quad (\varkappa = 1, 2, \ldots, N_e). \qquad (2b)$$

If we rewrite the Hamiltonian (I) into the new coordinates $\vec{S}, \ldots$ (regardless of the choice of the remaining coordinates of the particles) the kinetic-energy part splits up into two portions,

$$T(R) + T(\xi) = T(S) + \left\{ T'(R') + T'(\xi') + \ldots \right\}, \quad (3)$$

one of them, $T(S)$, describing the center-of-mass motion,

$$T(S) = \vec{p}_S^{\,2}/2M \qquad (4)$$

in the classical case, or

$$\hat{T}(S) = -(\hbar^2/2M)\vec{\nabla}_S^{\,2} \qquad (4')$$

in the quantum-mechanical formalism (see appendix). The potential energy is a function of the interparticle distances only and does not depend on the center-of-mass coordinates.

---

[1] The coordinates of the N-th nucleus relative to the center of mass can be calculated from the primed coordinates (2a,b).

Because of this form of the Hamiltonian, the center-of-mass motion can be readily obtained. Classically, the Hamiltonian equations of motion give $\dot{\vec{S}} = \vec{P}_S / M$ and $\dot{\vec{P}}_S = 0$, consequently $\ddot{\vec{S}} = $ const, i. e. the center of mass of the (isolated) system is moving rectilinear with uniform velocity. Analogously, in the quantum-mechanical treatment the Schrödinger equation is separable in the CM and remaining coordinates; the corresponding product ansatz for the total wavefunction leads to a solution $\Theta(\vec{S})$ in form of a plane wave for the CM part.

Summarizing these considerations, the problem of CM motion as that of a force-free mass point (mass M representing the isolated total system) can be exactly solved and left out of account henceforth. The dimension of the whole problem is thereby reduced by three degrees of freedom. The following considerations are concerned only with the internal motions of the system described by the internal coordinates as given, e. g., by eqs. (2) and determined by the interaction potential $U = U(R) + U(\xi, R) + U(\xi)$ in the classical Hamiltonian function

$$H_{int} = H - (\vec{P}_S^2 / 2M) \tag{5}$$

or in the quantum-mechanical Hamiltonian operator

$$\hat{H}_{int} = \hat{H} - (-\hbar^2/2M)\vec{\nabla}_S^2 . \tag{5'}$$

This internal motion provides all physically significant information about the collision processes. Of course, the motion in the different coordinate systems looks quite different, e. g. the scattering angles (ϑ, φ) have different values and, consequently, the differential cross sections have different forms. From the transformation relations of coordinates follow corresponding transformation relations for the angles which can be used for conversion of differential cross sections /4/. The total cross sections are invariant against changes of the coordinate system.

2.2. Separation of Electronic and Nuclear Motions. Interaction Potentials (Potential-Energy Surfaces)

2.2.1. Heuristic Considerations

In the theoretical treatment of elementary collision processes between two molecules P and Q , the problem consists essentially in following the motion of the nuclei of the constituent atoms including the attached electron clouds. In an heuristic way, we discuss here the origin of the forces determining this motion.

At large distances $P - Q$ there is no interaction between the collision partners; the free particles P and Q are characterized by notation of their internal states (electronic, vibration, rotation). In the course of the approach at first the outer parts of the electron clouds begin to interact, the electrons of P get more and more into influence of the nuclei of Q and vice versa; a common electron cloud of the total system $\{P, Q\}$ being formed. During this process the electron distribution and consequently the forces on the nuclei exerted by the electron cloud change continuously; the resulting electronic and nuclear forces acting between the constituents of $\{P, Q\}$ can be of attractive or repulsive nature.

If we make use of the approximate separability of electronic and nuclear motions as a consequence of the small mass ratio m_e/m_k mentioned above, we can arrive at an expression for the potential governing the motion of the nuclei - or, roughly speaking, the motion of the atoms $A, B, C, \ldots$ In the adiabatic approximation, the electron cloud is at any time in one and the same stationary quantum state corresponding to the momentary nuclear arrangement (and to the states of the asymptotically non-interacting particles P, Q).

Let us denote the energy of the electrons in the field of a fixed nuclear framework (characterized by the internuclear distances R_{AB}, ... , e. g.) by $\epsilon_n(R)$ depending parametrically on $R \equiv \{R_{AB}, ...\}$. Under the assumption that the quantum state of the electron cloud does not change during the collision process in the whole range of nuclear configurations passed through, we can state the following energy balance: the total energy E of the system is composed simply of the nuclear kinetic energy $T(R)$, the electrostatic nuclear repulsion energy $U(R)$, eq. (Ic), and the total electronic energy $\epsilon_n(R)$:

$$E = T(R) + U(R) + \epsilon_n(R). \qquad (1)$$

Taking this relation as the energy of the aggregate of atoms $\{P, Q\} \equiv \{A, B, C, ...\}$, averaged over many periods of the electronic motion, one obtains as the potential energy of the motion of the constituent atomic nuclei the expression

$$V_n(R) = U(R) + \epsilon_n(R), \qquad (2)$$

i. e. in this adiabatic approach the electronic energy represents the average potential of the forces exerted onto the nuclei by the electrons. A rigorous derivation as well as a discussion of the validity of this approach will be given in section 2.2.2.

In the most simple case of two interacting atoms, $\{P, Q\} = \{A, B\}$, the potential V is a function of the internuclear distance $R \equiv R_{AB}$; fig. 2.-1 shows two typical potential curves for non-bonding and bonding interactions, respectively. Potential functions of polyatomic systems ($N > 2$) show a much more complicated behaviour depending on $3N-6$ variables (internuclear distances, angles)

Interaction of two H atoms

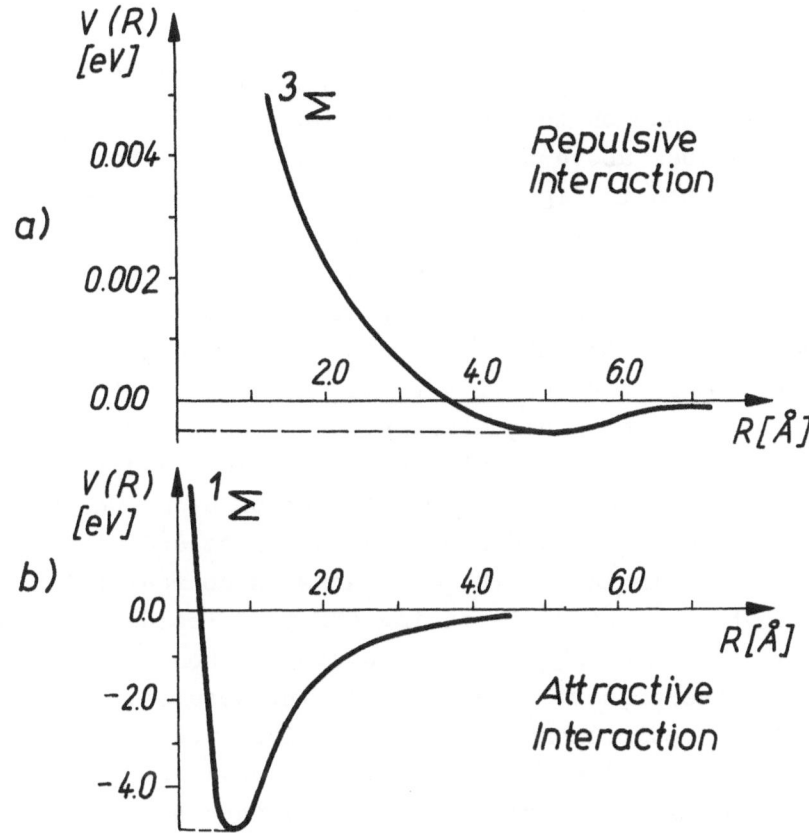

Fig. 2.-1

which characterize the relative positions of the nuclei.
Geometrically, such function represents a hypersurface in
a $(3N-5)$-dimensional space - the so-called "potential-
energy hypersurface" or simply "potential-energy surface".

According to the adiabatic approximation, eq. (2), to each quantum state n of the electron cloud belongs one such potential-energy surface.

For three nuclei moving along a straight line $A-B-C$ the potential depends on the two distances R_{AB} and R_{BC}; the potential values above the (R_{AB}, R_{BC}) plane form a surface as presented in fig. 2.-2 and, by means of a contour diagram, in fig. 2.-3. At large distances $A-BC$ we have a free molecule BC; a cut perpendicular to the R_{AB} axis gives a curve of the form 2.-1b. Approaching atom A, the interaction with molecule BC sets on and the potential curve $V_{mol}(R_{BC})$ becomes distorted. In the case of figs. 2.-2 and 2.-3 the minimum energy path (dashed-dotted line) rises up to a saddle point; in this region, all three atoms are near each other. Such arrangement of atoms we call "transition complex" regardless of its stability and life-time. Further diminishing of the distances is energetically unfavourable, the potential is rising steeply. Besides elongation of the distance R_{AB} (corresponding to non-reactive processes), in the present case also the increase of the distance R_{BC} (corresponding to a reactive exchange process $A + BC \longrightarrow AB + C$) is energetically preferred; the potential energy decreases. At large distances R_{BC} a cut through the surface perpendicular to the R_{BC} axis gives the potential curve of the free molecule AB. If the potential minimum of AB is higher than that of BC [1] or (and) if the potential surface shows a barrier (saddle point) between the "reactant valley" ($A+BC$) and the "product valley" ($AB + C$), the above-mentioned exchange process can proceed only in case the system possesses enough energy (e. g. relative translational energy of the partners $A-BC$

---

[1] In this case the potential is called "endoergic", in the reverse case "exoergic".

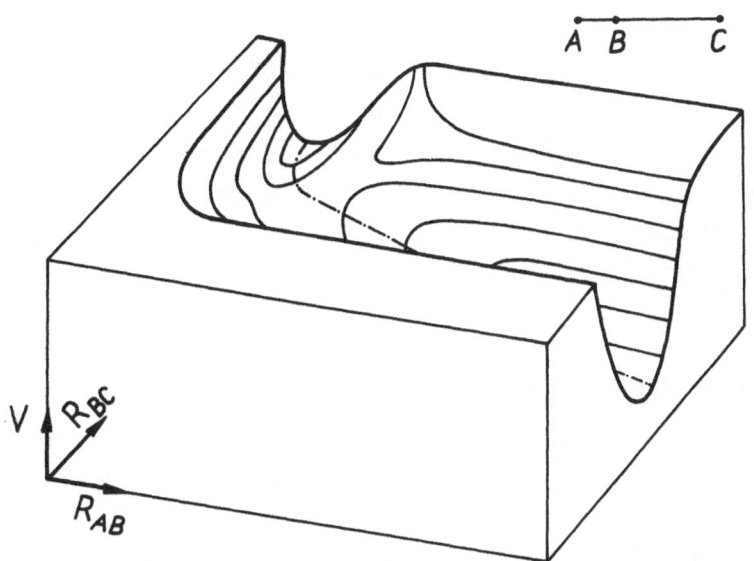

Fig. 2.-2

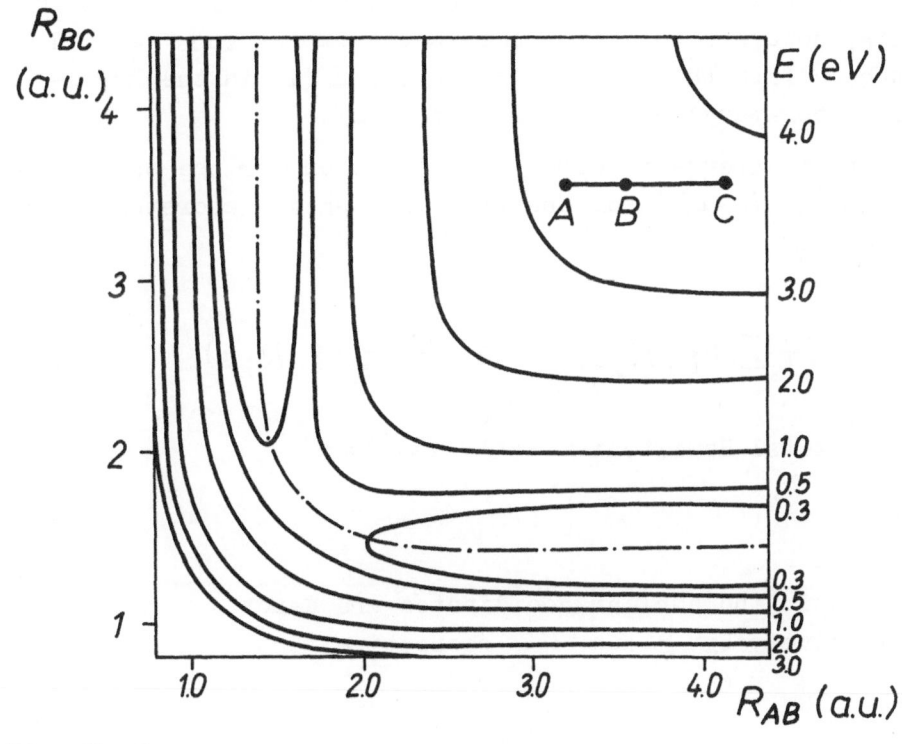

Fig. 2.-3

or internal energy of BC) to surmount the potential difference. If the system is so rich in energy that nuclear arrangements with R_{AB} as well as R_{BC} large can be realized, dissociation $A + BC \rightarrow A + B + C$ is possible.

For non-linear configurations of the three nuclei and for
systems with more than three nuclei, the potential-energy
function cannot be visualized as a whole.

2.2.2. <u>Born-Oppenheimer Separation. Adiabatic Approximation</u>

According to the heuristic considerations in the preceding
section, electronic and nuclear motions can be treated separately in an approximate way because of the small mass ratio
m_e/m_k resulting in order-of-magnitude differences of the
velocities of electrons and nuclei. A consequent formulation
of this approach should be based on the Hamiltonian $\hat{H}_{int}$,
after separation of the center-of-mass motion /3b/. Instead
of this, however, to avoid complicated expressions we will
start here from the original Hamiltonian (I) in the space-
fixed (laboratory) coordinate system /2/.

The time-dependent Schrödinger equation for the wavefunction $\Xi (\xi, R ; t)$[1] of the total system in laboratory coordinates reads as follows:

$$\hat{H} \, \Xi(\xi, R ; t) = i\hbar \, \frac{\partial}{\partial t} \, \Xi(\xi, R ; t) \qquad (3)$$

with the total Hamiltonian, eqs. (I, Ia-e),

---

[1]Electronic and nuclear coordinates are denoted
collectively by ξ and R , respectively.

$$\hat{H}(\xi, R) = \hat{T}(R) + \hat{H}_{el}(\xi, R) \tag{4}$$

where the terms collected in

$$\cdot\hat{H}_{el}(\xi, R) \equiv \hat{T}(\xi) + U(\xi, R) + U(\xi) + U(R) \tag{5}$$

represent the usual electronic Hamiltonian with the addition of the nuclear repulsion (fixed-nuclei approximation of $\hat{H}$).

At first, we treat the electronic motion in the field of a rigid, space-fixed nuclear framework. Let us suppose the stationary states of the electron cloud, as solutions of the electronic Schrödinger equation

$$\hat{H}_{el}(\xi, R)\phi_n(\xi, R) = V_n(R)\phi_n(\xi, R), \tag{6}$$

to be known; both the electronic energies and wavefunctions depend parametrically on the nuclear coordinates. Furthermore, the spectrum of $\hat{H}_{el}$ is assumed to be purely discrete[1] and the eigenfunctions to be orthonormalized[2]:

$$\int \phi_n^*(\xi, R)\phi_{n'}(\xi, R)d\xi \equiv \langle \phi_n | \phi_{n'} \rangle = \delta_{nn'}. \tag{7}$$

---

[1] The generalization to a partly continuous spectrum is easily accomplished.

[2] Integration over electronic coordinates will be denoted throughout by brackets.

Since these eigenfunctions of $\hat{H}_{el}$ form a complete set, the total wavefunction Ξ can be expanded:

$$\Xi(\xi, R; t) = \sum_{n'} \phi_{n'}(\xi, R)\, \psi_{n'}(R; t). \qquad (8)$$

Inserting this series into the Schrödinger equation (3), carrying out the differentiations with respect to the nuclear coordinates (operator $\hat{T}(R)$), multiplying the equation from the left by ϕ_n^* , integrating with respect to the electronic coordinates and paying regard to the orthonormality (7), a set of coupled differential equations,

$$\left\{\hat{T}(R) + V_n(R)\right\} \psi_n + \sum_{n'} \hat{C}_{nn'}\, \psi_{n'} = i\hbar \frac{\partial}{\partial t}\, \psi_n$$

$$(n = 0, 1, 2, \ldots), \qquad (9)$$

results for the determination of the coefficient functions $\psi_n(R; t)$ describing the nuclear motion. The coupling operators $\hat{C}_{nn'}$ are defined by

$$\hat{C}_{nn'} \equiv \langle \phi_n | \hat{T}(R) | \phi_{n'} \rangle - \sum_k \frac{\hbar^2}{m_k} \langle \phi_n | \vec{\nabla}_k | \phi_{n'} \rangle \vec{\nabla}_k \quad (9a)$$

and the (adiabatic) potential functions governing the nuclear motion are given by the eigenvalues of the electronic Schrödinger equation (6); the diagonal term $\hat{C}_{nn}$ of expression (9a) represents a correction to V_n .

On principle, the formulation of the Born-Oppenheimer treatment is not necessarily to be based on the adiabatic electronic states $\phi_n(\xi, R)$ as solutions of the complete electronic Schrödinger equation (6) but may be performed using some other orthonormal complete set of electronic wavefunctions depending parametrically on the nuclear coordinates

R . The general structure of the resulting set of nuclear wave equations is the same as that of eqs. (9).

Leaving aside the discussion of the coupling operators $\hat{C}_{nn'}$ we note that in case they are negligible and also the diagonal terms can be omitted, one gets a set of un-coupled nuclear Schrödinger equations,

$$\left\{ \hat{T}(R) + V_n(R) \right\} \psi_n = i\hbar \frac{\partial}{\partial t} \psi_n$$

$$(n = 0,1,2,\ldots), \qquad (10)$$

each for the different electronic states. This means, the nu-clear motion proceeds without changes of the quantum state of the electron cloud and, correspondingly, the wavefunction is reduced to a single term:

$$\Xi(\xi, R; t) \approx \phi_n(\xi, R) \psi_n(R; t) . \qquad (11)$$

This is the so-called adiabatic approximation which corresponds to the heuristic considerations in section 2.2.1.

If the coupling between different electronic states cannot be neglected, the processes are called non-adiabatic. In the course of a collision of such kind the electronic state of the system can change, and the theoretical treatment becomes in general extremely involved (see chapter 5.).

After separation of nuclear and electronic motion in the space-fixed coordinate system (laboratory system) as done here, in the usual procedure the center-of-mass motion of the nuclei only is separated subsequently. In this way, the (small but finite) influence of the electronic masses upon the nuclear motion is disregarded. The error caused by this approach, to be sure, is not momentous compared with the approximations introduced in the next steps of the theory (see the subsequent chapters). The rigorous treatment, performing Born-Oppenheimer separation after the center-of-mass separation described in appendix A 1., is quite analogous leading merely to more involved calculations and expressions /3b/; all general conclusions drawn above remain valid.

Let us now briefly discuss the conditions under which the concept of adiabatic separation of electronic and nuclear motions is justified. To this end, we simplify the problem further by using a so-called semiclassical approach (see section 4.2.) in which the "slow subsystem" (nuclei) is treated classically and the "fast subsystem" (electrons) quantum-mechanically: the motion of the nuclei is described by some (multi-dimensional) curve $R[t]$ in the nuclear configuration space,

$$R = R[t] , \tag{12}$$

and the electronic motion by some wavefunction Φ .

In the adiabatic approximation, the electronic state is given by the stationary solutions $\phi_n(\xi, R)$ of the electronic Schrödinger equation (6) depending parametrically on the nuclear coordinates R . Now, these coordinates are functions of time and the electronic Hamiltonian $\hat{H}_{el}$ becomes time-dependent. As a consequence, the electronic motion will be described by a non-stationary time-dependent wavefunction $\Phi(\xi; t)$ satisfying the time-dependent electronic Schrödinger equation,

$$\hat{H}_{el}(\xi, R[t]) \, \Phi(\xi; t) = i\hbar \frac{\partial}{\partial t} \, \Phi(\xi; t). \tag{13}$$

The solution $\Phi(\xi; t)$ can be represented by an expansion in terms of the adiabatic eigenfunctions $\phi_n(\xi, R)$; we take it in the form

$$\Phi(\xi; t) = \sum_n a_n(t) \, \phi_n(\xi, R) \exp\left\{ -\frac{i}{\hbar} \int^t V_n(R) \, dt \right\} \tag{14}$$

where the R-dependent quantities are functions of time according to eq. (12).

The perturbation of the adiabatic behaviour of the electron cloud (i.e. the instantaneous adjustment to the momentary nuclear configuration) caused by the motion of the nuclei is manifested now in the time-dependence of $\hat{H}_{el}$ and $\hat{\Phi}$; the square modulus of the expansion coefficients, $|a_n(t)|^2$, being a function of time gives the probability of finding the system in the adiabatic electronic state n . In this way the nuclear motion effects "non-adiabatic transitions" between the various adiabatic electronic states. Equations for the determination of the coefficients $a_n(t)$ are obtained by insertion of expansion (14) into the Schrödinger equation (13) utilizing eqs. (6) and (7):

$$i\hbar\,\dot{a}_n = \sum_{n'(\neq n)} C_{nn'}\exp\left\{-\frac{i}{\hbar}\int^t (V_{n'}-V_n)\,dt\right\} a_{n'} \qquad (15)$$

with the coupling matrix elements

$$C_{nn'} \equiv \langle\,\phi_n\,|-i\hbar\,\partial/\partial t\,|\,\phi_{n'}\,\rangle\,. \qquad (15a)$$

The expansion (14), the set of equations (15) and the coupling matrix elements (15a) of the semiclassical approach are analogous to the expansion (8), the set of equations (9) and the coupling operators (9a), respectively, of the rigorous quantum-mechanical formulation.

A study of this set of equations enables us to formulate conditions under which the coefficients a_n may be considered as being approximately constant - in other words, conditions for the validity of the adiabatic approximation. Let be $\Delta V(R)$ the difference of two arbitrary adiabatic terms at point R of the nuclear configuration space, $l(R)$ a characteristic length over which the corresponding electronic wavefunctions ϕ substantially change, and u some average(classical) velocity of the nuclei at point R . Then the quantity

$$\gamma(R) = \frac{\Delta V \cdot l}{\hbar u} \, , \qquad (16)$$

the so-called Massey parameter, gives the ratio of the passage time of the nuclei through the region of dimension l around point R to the time $\hbar / \Delta V$ characterizing the electronic motion being just the inverse transition frequency between the two adiabatic states. In other words, γ represents the ratio of the time the perturbation by the nuclear motion is acting to the period of the internal motion of the electronic subsystem. This definition is, of course, a very rough one because the perturbation changes the eigenfrequencies; it is adequate only in cases where these changes are small.

It is plausible (and can be shown more rigorously, see chapter 5.) that for large ΔV and small u ($\gamma \gg 1$), non-adiabatic transitions will occur with small probability; the electron cloud has time to adjust its state instantaneously to the slowly changing nuclear configuration. In those regions of the nuclear configuration space where the Massey parameter is not large (i.e. where adiabatic potential surfaces are close together or intersect, where electronic wavefunctions are changing very rapidly for varying nuclear coordinates, where nuclei are moving with high velocity), non-adiabatic transitions become probable, $V(R)$ loses its meaning as a potential energy, and the coupling between both subsystems (electronic and nuclear) must be taken into account, e.g., by solving eqs. (15) or even eqs. (9). This problem will be further pursued in chapter 5.

It should be noted that we did not discuss here the determination of the trajectories (12) of the classical subsystem. For the adiabatic case this will be considered in section 3.1.; since then the electronic motion does not explicitly occur at all, this treatment is called the classical approach. Furthermore, in section 4.2. we briefly present a more rigorous formulation of the semiclassical approach taking into account also the influence of the perturbed electronic motion back on the (classically calculated) nuclear motion. Finally, we point out that the separation of the total system into a "slow" and a "fast" subsystem does not mean necessarily nuclear and electronic motion, respectively, but may also relate to modes of nuclear motion differing substantially in characteristic velocities.

2.2.3. Present State of Potential-Energy-Surface Calculations

According to the considerations in the two preceding sections, for determining the adiabatic molecular interaction potentials, $V_n(R)$, the electronic Schrödinger equation (6) must be solved. Without going into the details, we merely indicate here the basic difficulties connected with the problem and illustrate the recent progress; for a more thorough discussion of methods and results the reader is referred to the cited literature.

Apart from simple models, the potential energy $V_n(R)$ as a function of the nuclear coordinates R cannot be obtained directly in an analytic form; therefore it is necessary to calculate numerically approximate solutions of eq. (6) for a number of discrete nuclear configurations in the range which is energetically accessible to the system considered for a certain total energy. The number of these configurations must be sufficiently large to make a subsequent analytic fitting procedure possible. Furthermore, the accuracy required for an adequate description of the nuclear motion is high: about 1 kcal/mole in "critical regions"[1] (so-called

---

[1] For example, in the neighbourhood of potential barriers.

chemical accuracy). As a rule, the independent-particle (Hartree-Fock) approximation /5/ is not sufficient and electronic correlation /5/ must be taken into account properly.

These very severe requirements cannot be fulfilled by applying one of the conventional semiempirical methods /6/ being in use at present; in some ways, the DIM (Diatomics in Molecules) approach /7/ seems to be promising. Quantitatively reliable results, on principle, are to be expected from appropriate ab-initio methods /5/ - but with enormous expenses. In table 2.-1 some representative examples of calculations are collected to demonstrate the present situation; most of them are restricted to the electronic ground state.

Notwithstanding the progress during the last decade /8/, it can be inferred from these data that we should not count on accurate complete ab-initio potential-energy surfaces for medium-sized and large molecules of chemical interest before long, if at all. Most probably, however, this will be not necessary actually: chemical reactions take place usually at definite parts of a molecule so that one has to describe only these parts and their interactions with attacking partners very accurately, using a lump description of the remaining degrees of freedom. Furthermore, there is some hope to find appropriate combined methods giving in the first step (say, semiempirically) an analytic potential of correct global behaviour with some free parameters; after having performed very accurate ab-initio calculations for a small number of carefully selected critical nuclear configurations, the parameters are then adjusted to fit these points. Finally, it should be pointed out that the above-indicated accuracy is not always required; it can be renounced, e.g., frequently in calculations of high-temperature rate constants.

Table 2.-1: Ab-initio calculations of potential-energy
surfaces (selected examples)

N	N_e	System	Ref.	Approach	Accuracy
3	2...4	H_3^+	a	incl. electron correlat.	chemical for linear configurat.
		H_3	b		
		HeH_2^+	c		
		LiH_2^+	d		
3	$\lesssim 20$	FH_2	e	incl. electron correlat.	$\leqslant$ chemical for linear configurat.
		LiN_2^+	f		
4	$\leqq 4$	H_4	g		
> 3	> 4	FCH_3F^-	h	SCF	qualitative
		FCH_3CN^-	i		

a   BAUSCHLICHER et al., J. Chem. Phys. 59 (1973) 1286.

b   LIU, J. Chem. Phys. 58 (1973) 1925.

c   BROWN/HAYES, J. Chem. Phys. 55 (1971) 922;
EDMISTON et al., J. Chem. Phys. 52 (1970) 3414.

d   KUTZELNIGG et al., Chem. Phys. 1 (1973) 27.

e   BENDER et al., J. Chem. Phys. 56 (1972) 4626.

f   STAEMMLER, Chem. Phys. 7 (1975) 17.

g   SILVER/STEVENS, J. Chem. Phys. 59 (1973) 3378.

h   DEDIEU/VEILLARD, J. Am. Chem. Soc. 94 (1972) 6730.

i   DUKE/BADER, Chem. Phys. Letters 10 (1971) 631.

In dealing with non-adiabatic processes, besides the adiabatic potential-energy functions, $V_n(R)$, or some proper approximations to them, the corresponding coupling terms $C_{nn'}$ must be calculated (see section 2.2.2.). In general, this requires the explicit evaluation of the electronic wavefunctions, $\phi_n(\xi, R)$, to a sufficiently high accuracy. It should be noticed that this difficult problem[1] sometimes can be circumvented, for example, in the above-mentioned DIM approach.

2.3. Scattering Channels

For systematic characterization of the different elementary processes which can occur in the course of a collision $P + Q$ the concept of scattering channel (see /9/, e. g.) proves useful. A scattering channel is a possible decomposition of the total system $\{A, B, C, ...\}$ into stable subsystems being in definite quantum states i, j , etc. Quantities related to specific scattering channels will be designated by Greek letters α , β , ... henceforth.

As examples let us consider the processes (1.-2). The separated reactants $P(i) + Q(j)$ represent the entrance channel (elastic channel), the products $P(i') + Q(j')$ give the diverse inelastic channels, and the various kinds of products $X(l) + Y(m) + ...$ correspond to reaction channels.

By reason of energy conservation, a collision of two reactants P and Q in definite internal states i and j , respectively, (entrance channel α) proceeding with definite

---

[1] According to a well-known theorem of quantum mechanics (see, e.g., /5,9/), variationally determined wavefunctions usually are much less accurate than the corresponding energies.

collision energy E_{tr}^{α} [1] can lead to a limited number of processes only. The total energy E of the system is given by

$$E = E_{tr}^{\alpha} + E_{int}^{\alpha} \tag{1}$$

where E_{int}^{α} represents the sum of the reactant internal energies, $E_{int}^{\alpha} \equiv E_{int}(P(i)) + E_{int}(Q(j))$.
For any process leading to product channel β, i. e. molecules X and Y in internal states l and m, respectively, the energy conservation law demands

$$E_{tr}^{\beta} + E_{int}^{\beta} = E \tag{1'}$$

with $E_{int}^{\beta} \equiv E_{int}(X(l)) + E_{int}(Y(m))$. This process can be realized only if

$$E_{int}^{\beta} < E \tag{2}$$

$(E_{tr} > 0)$. All channels which satisfy this condition are called "open channels", all others are "closed channels".

2.4. Classification of Elementary Processes. Microscopic Mechanism

In the course of a collision process between two molecules P and Q all constituents of the system will stay for a certain time close together in a region of molecular size (several Angstroms). This arrangement of atoms is called "collision complex" or "transition complex"; after some time it decays into products:

---

[1] The relative translational energy E_{tr}^{α} is given by $\mu_{\alpha} u_{\alpha}^2/2$ where $\mu_{\alpha} \equiv m_P m_Q /(m_P + m_Q)$ is the reduced mass of the particle pair $P - Q$ and u_{α} denotes the relative velocity $P - Q$.

$$P + Q \rightarrow [PQ] \rightarrow X + Y . \tag{1}$$

If the lifetime τ of this collision complex $[PQ]$ is longer
than the period of the slowest forms of motion in such quasi-
molecular aggregate (overall and internal rotations), i.e.

$$\tau > \text{several } 10^{-12} \ldots 10^{-11} \text{s}, \tag{2}$$

then, in consequence of the interactions, an energy exchange
among the internal degrees of freedom of the complex can
take place. In this case, the further course of the process
will be determined mainly by the properties of the collision
complex, by the statistical distribution of the energy over
the degrees of freedom /10a/, and the theoretical treatment
can be considerably simplified by application of statistical
methods.

The occurrence of long-lived collision complexes is more
probable for systems with many degrees of freedom than for
those with few degrees of freedom. The first who succeeded
in proving indirectly the existence of such complexes was
HERSCHBACH 1967 using the molecular-beam method in investigations of reactive elementary processes $Cs + RbCl$
$\rightarrow CsCl + Rb$. The differential cross section of processes passing via a long-lived complex shows in the CM
coordinate system a symmetric form with respect to 90°
referred to the direction of relative motion (see fig.
2.-4a).

In so-called direct processes which pass via a short-
lived collision complex immediately from the reactants to
the products,

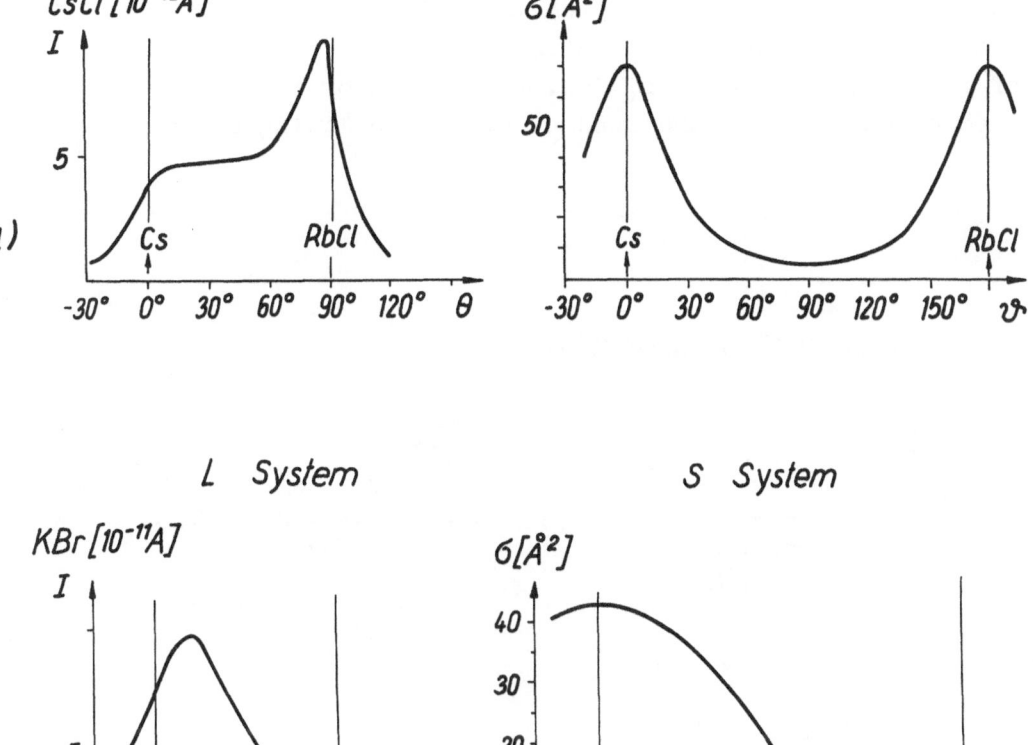

Fig. 2.-4

$$\tau \; < \; 10^{-12}\text{s}, \qquad\qquad (3)$$

the constituent particles "remember" their history; those cases on principle demand an investigation of the dynamics of the collision process, i.e. the motion of the particles themselves. This is the subject of the following chapter.

Rebound Mechanism

a)

Stripping Mechanism

b)

Fig. 2.-5

Differential cross sections for direct processes exhibit a
marked asymmetry (see fig. 2.-4b)[1] /10b/. Let us consider, in
particular, collisions of atoms A with diatomic molecules
BC ; several different prototypes of collision mechanisms
are possible. Reactive processes $K + HBr \longrightarrow KBr + H$,
for instance, are governed by the so-called rebound mechanism: the newly formed molecule (AB) is thrown back into
the reverse direction of the approaching atom (A) because
of strong repulsion forces between AB and C (see fig.
2.-5a). The differential reaction cross section in the CM
coordinate system shows a peak at scattering angles near
180° with respect to the direction of incoming A . The
stripping mechanism which prevails, e. g., in processes
$K + Br_2 \longrightarrow KBr + Br$, originates in the attractive
interaction between atoms A and B ; atom A pulls off atom
B in forward direction (see fig. 2.-5b) and, consequently,
the differential cross section exhibits a peak at 0° (see
fig. 2.-4b).

As a rule, mechanisms of this kind do not appear in pure
form; the course of the collision process is generally
determined in a very complicated way by the interaction
potential between the atoms as well as by the collision
energy E_{tr} and the internal states of the reactants.
The influence of the collision energy is immediately evident:
for increasing collision energy a complex mechanism will go
over into a direct one; furthermore, a collision proceeding
at low energies according to a rebound mechanism usually
will follow a stripping-type mechanism at higher energies.

---

[1] This statement represents a general rule although there
are exceptions.

3. Dynamics of Atomic and Molecular Collisions: Electronically Adiabatic Processes

After the presentation of general features of atomic and molecular collision theory in the preceding chapter, we now take up the consequent treatment of the dynamics restricting ourselves first to adiabatic processes with respect to electronic motion. Some more general concepts taking into account the coupling of electronic and nuclear motions are briefly discussed in chapters 4. and 5.

Neglecting the effect of the electron mass and presuming the adiabatic approximation as defined in sections 2.2.1. and 2.2.2., in a rigorous quantum-mechanical treatment the Schrödinger equation (2.2.-10) for the N nuclei,

$$\left\{ -\frac{\hbar^2}{2} \sum_{k=1}^{N} \frac{1}{m_k} \vec{\nabla}_k^2 + V(\vec{R}_1,...,\vec{R}_N) \right\} \Psi = i\hbar \frac{\partial}{\partial t} \Psi, \quad (1)$$

is to be solved according to appropriate boundary and initial conditions (see refs. /9,13/ and section 3.2.) imposed on the wavefunction $\Psi \equiv \Psi(\vec{R}_1,...,\vec{R}_N;t)$. For more than two nuclei, this represents a very difficult problem which can be attacked nowadays, with reasonable expense, only by introducing serious approximations.

A considerable simplification would be to abandon the quantum-mechanical treatment and to describe the nuclear motion classically. In this case the nuclei are characterized by some spatial coordinates $q_i(t)$ and the conjugate generalized momenta $p_i(t)$ as dynamical variables resulting as solutions of the canonical equations of motion /5c,11/:

$$\dot{q}_i = \frac{\partial H}{\partial p_i} \quad , \quad \dot{p}_i = -\frac{\partial H}{\partial q_i} \quad , \quad (2)$$

$$(i = 1, 2, ..., 3N)$$

where $H \equiv H(q_1, \ldots, q_{3N}, p_1, \ldots, p_{3N})$ denotes the classical Hamiltonian. The merits of the classical method lie in its mathematical simplicity and physical visuality. Therefore, we discuss this approach in the following section 3.1. in detail whereas section 3.2. presents some principal aspects of quantum-mechanical methods.

3.1. Classical Approach

3.1.1. Some Arguments for the Reliability of the Classical Approach

By expansion of the wavefunction into powers of Planck's constant h , from the Schrödinger equation (1) follow formal criteria for the validity of classical mechanics (compare, e.g., ref. /9/). They are rather difficult to apply practically and will not be given here; we confine ourselves to some simple, rough estimates in analogy to optics. Geometrical optics considering paths of "light particles" holds if the optical properties of the medium do not change essentially over regions of extension of many wavelengths. Correspondingly, classical mechanics is valid if the potential does not change appreciably over many deBroglie wavelengths.

Let us consider, for simplicity, the one-dimensional motion of a particle of mass m and energy E in a potential V , a being a characteristic length over which the potential can be supposed to be approximately constant. For potentials of molecular chemical interactions, a has the order of magnitude of several Angstroms, except the region of the transition complex. The local kinetic energy of the particle is $E_{tr} = E - V$ and the local deBroglie wavelength is given by

$$\lambda = h/p = h/\sqrt{2m(E-V)} . \qquad (3)$$

The condition for validity of classical mechanics,

$$\lambda \ll a , \qquad (4)$$

is fulfilled the better the larger is E and the heavier is the particle.

As the hitherto existing experience shows, classical mechanics gives qualitatively and partly quantitatively correct results even for light nuclei as H, D and He. On principle, of course, one has to expect classical calculations to fail in certain specific aspects of the collision process as, for instance, tunneling and resonance phenomena (see section 3.2.).

3.1.2. <u>Atom-Atom Collisions. Elastic Scattering</u>

At the beginning, we demonstrate some important features of the classical approach by considering the simple case of two-particle scattering.

The two particles A and B (atomic nuclei) are treated as point masses interacting by a potential $V(R)$ which depends only on the distance R between A and B. The number of (nuclear) degrees of freedom of the system is $f = 6$; by separation of the center-of-mass motion[1] f can be reduced to 3. To this end, we define the coordinates of the nuclear center of mass,

---

[1] The effect of the electron masses is neglected.

$$\vec{S}_n = (m_A \vec{R}_A + m_B \vec{R}_B)/(m_A + m_B), \qquad (5)$$

and the coordinates of A relative to B,

$$\vec{R} = \vec{R}_A - \vec{R}_B . \qquad (6)$$

Using these coordinates, the Lagrangian function of the system takes the form

$$L' \equiv T - V = \frac{m_A + m_B}{2} \dot{\vec{S}}_n^{\,2} + \frac{\mu}{2} \dot{\vec{R}}^{\,2} - V(R), \qquad (7)$$

and the angular momentum is

$$\vec{\ell}' = (m_A + m_B)(\vec{S}_n \times \dot{\vec{S}}_n) + \mu (\vec{R} \times \dot{\vec{R}}) \qquad (8)$$

where μ denotes the effective mass

$$\mu \equiv m_A m_B / (m_A + m_B) \qquad (9)$$

of the particle pair $A - B$. To eliminate the CM motion we formally put $\vec{S}_n = 0$ and consider henceforth only the parts $L \equiv L' - (m_A + m_B) \dot{\vec{S}}_n / 2$ and $\vec{\ell} \equiv \vec{\ell}' - (m_A + m_B)(\vec{S}_n \times \dot{\vec{S}}_n)$ related to the internal motion of the system with $f = 3$ degrees of freedom.

It can be easily shown that, because of the spherical symmetry of the potential, the internal angular momentum $\vec{\ell}$ is time-independent[1]. Consequently, the motion proceeds in a fixed plane perpendicular to $\vec{\ell}$; this plane we choose as the (x, y) plane reducing thereby the problem additionally to $f = 2$ degrees of freedom. By rewriting in planar polar coordinates R and χ ,

$$x = R \cos\chi \;, \quad y = R \sin\chi \;, \tag{10}$$

the Lagrangian function is given as

$$L = \frac{\mu}{2}(\dot{R}^2 + R^2\dot{\chi}^2) - V(R). \tag{11}$$

Instead of applying the Newtonian or Lagrangian formalism for solution of the simple two-particle problem we use for didactic reasons the Hamiltonian formalism which is the most appropriate one in cases of more than two particles. We have to introduce the generalized momenta p_R and p_χ conjugate to the coordinates R and χ , respectively,

$$p_R \equiv \frac{\partial L}{\partial \dot{R}} = \mu\dot{R} \;, \quad p_\chi \equiv \frac{\partial L}{\partial \dot{\chi}} = \mu R^2\dot{\chi} \;, \tag{12}$$

and to substitute them into the Hamiltonian function:

---

[1] This is seen most simply by calculating $d\vec{\ell}/dt$ and taking into account Newton's equation of motion $\mu\ddot{\vec{R}}$ $= -\text{grad } V(R) = -(\partial V/\partial R)(\vec{R}/R)$.

$$H \equiv H(R, \chi, p_R, p_\chi) \equiv T + V$$
$$= (p_R^2/2\mu) + (p_\chi^2/2\mu R^2) + V(R). \qquad (13)$$

The canonical equations of motion (2) represent a set of four ordinary first-order differential equations:

$$\dot{p}_R = -\frac{\partial H}{\partial R} = (p_\chi^2/\mu R^3) - (\partial V/\partial R), \qquad (14a)$$

$$\dot{p}_\chi = -\frac{\partial H}{\partial \chi} = 0, \qquad (14b)$$

$$\dot{R} = \frac{\partial H}{\partial p_R} = p_R/\mu, \qquad (14c)$$

$$\dot{\chi} = \frac{\partial H}{\partial p_\chi} = p_\chi/\mu R^2. \qquad (14d)$$

Solution of the Equations of Motion

In principle, from these equations (14a–d) the four functions $R(t)$, $\chi(t)$, $p_R(t)$ and $p_\chi(t)$ can be calculated. We follow here, however, a simpler way using the conservation of energy and angular momentum written in the form:

$$H = \frac{\mu}{2}(\dot{R}^2 + R^2\dot{\chi}^2) + V(R) = \text{const} \equiv E, \qquad (15a)$$

$$|\vec{\ell}| = \mu R^2 \dot{\chi} = p_\chi = \text{const} \equiv \ell. \qquad (15b)$$

At the beginning we have to specify the initial conditions for the motion; the necessary quantities are defined in fig. 3.–1 showing a trajectory of the particle A relative

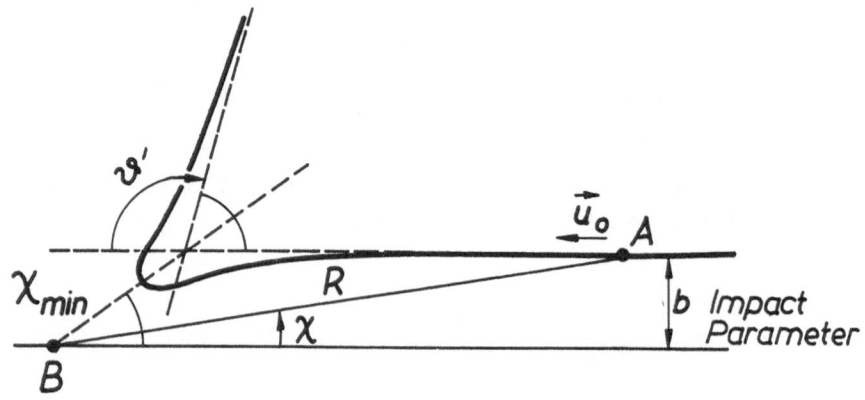

Fig. 3.-1

to particle B . The initial state is characterized by the
impact parameter b and the relative velocity u_o of the
freely moving particle A (i.e., at large distances R ,
without interaction). The angle of deflection, ϑ', is related to the polar angle χ_{min} of the position of minimal
distance R_{min} between A and B by

$$\vartheta' = \pi - 2\chi_{min} . \qquad (16)$$

Using the constants of motion

$$E = \frac{\mu}{2} u_o^2 \quad , \quad \ell = \mu b u_o \qquad (17)$$

and rewriting the expressions (15a,b) one arrives at the
following equations of motion which are equivalent to the

Hamiltonian equations (14a–d):

$$-\frac{dR}{dt} = \left\{ \frac{2E}{\mu} - \frac{\ell^2}{\mu R^2} - \frac{2V(R)}{\mu} \right\}^{1/2} = u_o \left\{ 1 - \left(\frac{b}{R}\right)^2 - \frac{V(R)}{E} \right\}^{1/2}, \quad \text{(18a)}$$

$$\frac{d\chi}{dt} = \frac{b u_o}{R^2}. \quad \text{(18b)}$$

From these relations the functions $R(t)$ and $\chi(t)$ describing the motion can be calculated. We are interested in the trajectory followed by A relative to the position of B ; it is determined by the equation

$$\frac{d\chi}{dR} = \frac{d\chi}{dt} \bigg/ \frac{dR}{dt} = -\frac{b}{R^2} \left[1 - \left(\frac{b}{R}\right)^2 - \frac{V(R)}{E} \right]^{-1/2} \quad \text{(19)}$$

and can be evaluated as the integral

$$\chi(R) = \int_R^\infty dR' \, b \bigg/ \left(R'^2 \left[1 - (b/R')^2 - V(R')/E \right]^{1/2} \right). \quad \text{(20)}$$

The deflection angle ϑ' , according to eq. (16), is equal to $\pi - 2\chi(R_{min})$ with the distance R_{min} of nearest approach given by the condition $\dot{R} = 0$, i.e. the largest positive root of the equation

$$\frac{\ell^2}{2\mu R^2} + V(R) = E, \quad \text{(21)}$$

see eq. (18a). The quantity ϑ' depends on the parameters b and E ; for a fixed value of E the function $\vartheta'(b)$

is called deflection function[1].

Discussion of the Deflection Function

The deflection function is intimately connected with the form of the potential and leads immediately to the elastic differential cross section. For illustration we consider in fig. 3.-2 typical trajectories for a potential showing a minimum at $R = R_m$ (e.g., a Lennard-Jones potential); in fig. 3.-3 the corresponding deflection function is given. We introduce reduced quantities $\beta \equiv b/R_m$ and $\varrho \equiv R/R_m$ using R_m as unit length.

For small impact parameters (i.e. near-central collisions) the repulsive part of the potential is dominating and $\vartheta' > 0$; for $\beta = 0$ exact backward scattering occurs ($\vartheta' = \pi$). With increasing β the deflection decreases until attractive and repulsive forces cancel each other leading to non-deflected trajectories ($\vartheta' = 0$) at an impact parameter β_0. For $\beta > \beta_0$ the long-range attractive forces prevail, the deflection function becomes negative. At a certain value $\beta = \beta_r$ a minimum occurs, and for even higher impact parameters ϑ' goes monotonously to zero.

This form of the deflection function is characteristic for potentials with a minimum and for rather high collision energies E; at smaller energies the minimum of $\vartheta'(\beta)$ can degenerate into a singularity: $\vartheta' \to -\infty$ for $\beta \to \beta_r$. This phenomenon can be explained by considering the effective potential – the sum of the interaction potential $V(\varrho)$ and the centrifugal potential:

---

[1] In general, $\vartheta'(b)$ cannot be written in closed analytical form; the integral (20) is to be computed numerically.

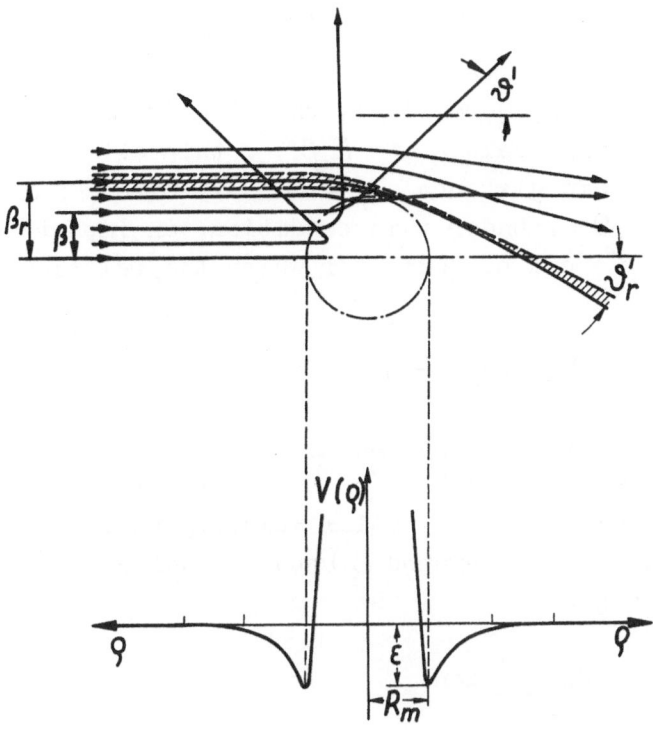

Fig. 3.-2

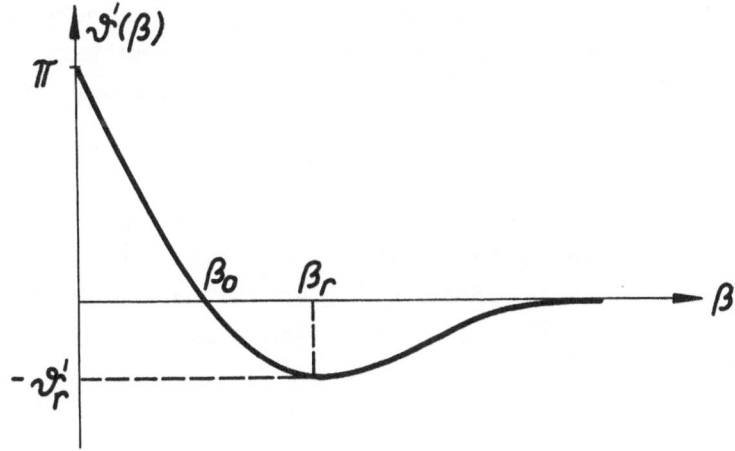

Fig. 3.-3

$$V_{eff}(\varrho) = V(\varrho) + E\beta^2/\varrho^2 . \qquad (22)$$

The latter is decreasing more slowly than the former for $\varrho \rightarrow \infty$; hence, for not too high values of $E\beta^2$, this effective potential at a certain distance $\varrho = \varrho'$ exhibits a maximum $V_{eff}^{\circ} \equiv V_{eff}(\varrho') > 0$. Under this condition, an impact-parameter value β° can be found for which the relation

$$V_{eff}(\varrho') = E \qquad (23)$$

holds so that the particle with energy E just is able to climb the "rotational barrier" $V_{eff}(\varrho')$. According to eq. (18a), in this case the radial velocity vanishes ("orbiting" of particle A around B) and the deflection function (16) diverges.

Calculation of CM Cross Section

Let us denote the current density of the incoming par-
ticles A in the CM coordinate system by j_o and let the impact parameter b correspond to the scattering angle ϑ (see fig. 3.-4)[1]. The particles with impact parameters lying in the interval $(b, b+db)$ are scattered into the angle interval $(\vartheta, \vartheta + d\vartheta)$; the number $d\mathcal{N}$ of these particles is given by (see chapter 1.)

---

[1] The physically measurable scattering angle ϑ (>0) is not identical with the deflection angle ϑ' ; the latter can have values $> \pi$ and < 0 . There holds the relation $\vartheta = |\vartheta'|$ modulo π .

43

$$d\mathcal{N} = j_o\,2\pi b\,db = \sigma(\vartheta)\,j_o\,2\pi \sin\vartheta\,d\vartheta \qquad (24)$$

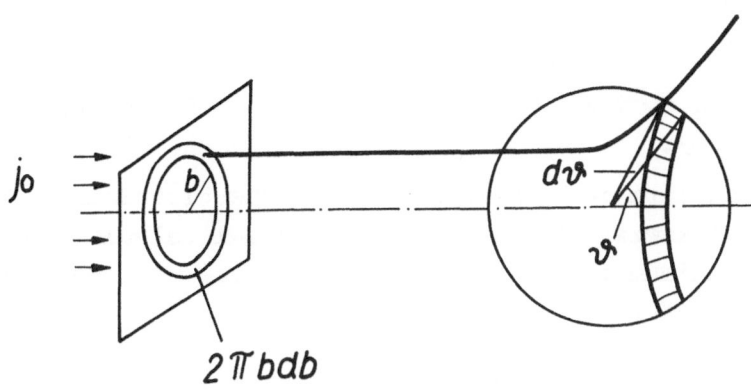

jo

$2\pi b\,db$

Fig. 3.-4

using the cylindrical symmetry of the problem. In the general
case, different impact parameters b_i can lead to the same
scattering angle ϑ (e.g., in fig. 3.-3 for $\vartheta < \vartheta_r$,
three impact parameters contribute to one value of ϑ);
therefore we get for the differential cross section:

$$\sigma(\vartheta) = \sum_i \frac{b_i}{\sin\vartheta}\left|\frac{db_i}{d\vartheta}\right| \qquad (25)$$

$(\sigma \geqslant 0)$.

The differential elastic cross section (25) becomes
singular $(\sigma \rightarrow \infty)$ for:

1) $d\vartheta/db = 0$, i.e. for $b = b_r$ in the example of fig. 3.-3.

This singularity represents a shortcoming of the classical approximation. It disappears in the quantum-mechanical treatment, instead a finite maximum occurs near ϑ_r ("rainbow scattering").

2) $\vartheta \rightarrow 0$, i.e. for $b = b_0$ and $b \rightarrow \infty$ in the example of fig. 3.-3.

These singularities are artifacts of the classical approach likewise. The quantum-mechanical calculation gives a maximum in forward direction, i.e. for small ϑ ("glory scattering").

Concerning further details we refer to the literature /12,13/.

3.1.3. Quasiclassical Treatment of Elementary Processes in Triatomic Systems: Inelastic and Reactive Scattering

In applying classical mechanics to elementary processes in triatomic systems $\{A, B, C\}$, e.g. collisions of an atom A with a diatomic molecule BC , we start again from the separation of electronic and nuclear motion in the laboratory coordinate system presuming the adiabatic approximation. The nuclear system has $f = 9$ degrees of freedom described, e.g., by the space-fixed Cartesian coordinates

$$\{\vec{R}_A, \vec{R}_B, \vec{R}_C\} \equiv \{\bar{q}_1, \ldots, \bar{q}_9\}. \qquad (26)$$

By separation of the CM motion the number of degrees of freedom reduces to $f = 6$. To this end we introduce nuclear

CM coordinates

$$\vec{S}_n = (m_A \vec{R}_A + m_B \vec{R}_B + m_C \vec{R}_C)/(m_A + m_B + m_C) \quad \text{(27a)}$$

(neglecting, as done before, the effect of the electron masses) and so-called channel-adapted relative coordinates[1]

$$\vec{r} = ' \vec{R}_C - \vec{R}_B , \quad \text{(27b)}$$

$$\vec{R} = \vec{R}_A - (m_B \vec{R}_B + m_C \vec{R}_C)/(m_B + m_C), \quad \text{(27c)}$$

as illustrated in fig. 3.-5.

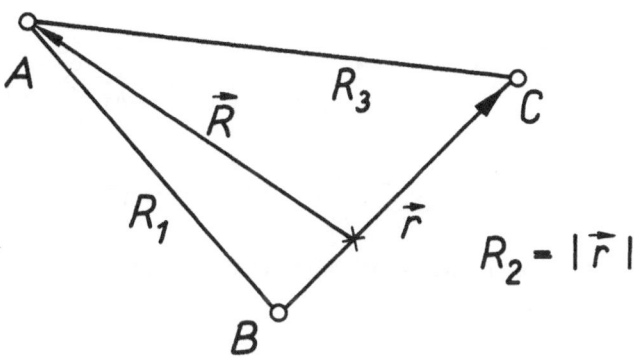

Fig. 3.-5

---

[1] These relative coordinates are appropriate for describing particle arrangements $A + BC$ in the entrance channel.

Equations of Motion and Initial Conditions

To write the equations of motion in a convenient form we denote the Cartesian components of the vectors, introduced in eqs. (27), and the corresponding conjugated momenta as follows:

$$\vec{r} \equiv \{q_1, q_2, q_3\} \quad , \quad \vec{R} \equiv \{q_4, q_5, q_6\} , \qquad (28a)$$
$$\vec{S}_n \equiv \{q_7, q_8, q_9\} ,$$

$$\vec{p}_r \equiv \{p_1, p_2, p_3\} \quad , \quad \vec{p}_R \equiv \{p_4, p_5, p_6\} , \qquad (28b)$$
$$\vec{p}_S \equiv \{p_7, p_8, p_9\} .$$

The complete Hamiltonian function H' has the form

$$H' = \frac{1}{2\mu_{BC}} \sum_{j=1}^{3} p_j^2 + \frac{1}{2\mu_{A,BC}} \sum_{j=4}^{6} p_j^2 + \frac{1}{2M} \sum_{j=7}^{9} p_j^2 \qquad (29)$$
$$+ V(R_{AB}(q_1, \ldots, q_6), \ldots)$$

where M is the total mass of the three nuclei,

$$M \equiv m_A + m_B + m_C , \qquad (29a)$$

and the reduced masses μ_{BC} and $\mu_{A,BC}$ are defined as

$$\mu_{BC} \equiv m_B m_C /(m_B + m_C) , \qquad (29b)$$

$$\mu_{A,BC} \equiv m_A (m_B + m_C)/M . \qquad (29c)$$

Since the potential depends only on the internal coordinates $q_1, \ldots, q_6$, the CM motion can be separated and the internal motion of the system is determined by the Hamiltonian function

$$H = \frac{1}{2\mu_{BC}} \sum_{j=1}^{3} p_j^2 + \frac{1}{2\mu_{A,BC}} \sum_{j=4}^{6} p_j^2$$
$$+ V(R_{AB}(q_1, \ldots, q_6), \ldots) .$$

(30)

The canonical equations of motion, eq. (2), more explicitly read as follows:

$$\dot{q}_j = p_j / \mu_{BC} \qquad\qquad (j = 1,2,3), \quad (31a)$$

$$\dot{q}_j = p_j / \mu_{A,BC} \qquad\qquad (j = 4,5,6), \quad (31b)$$

$$\dot{p}_j = - \sum_{i=1}^{3} (\partial V / \partial R_i)(\partial R_i / \partial q_j) \qquad (j = 1, \ldots, 6), \quad (31c)$$

denoting the internuclear distances R_{AB} , R_{BC} , R_{AC} by R_1 , R_2 , R_3 , respectively.

For calculation of the functions $q_1(t), \ldots, q_6(t)$ and $p_1(t), \ldots, p_6(t)$ from this system of 12 first-order ordinary differential equations, we have to specify the initial conditions, i.e., the values of q_j and p_j at some fixed time chosen so that atom A and molecule BC are far apart from each other (without interaction).

As explained in fig. 3.-6, the origin of the coordinate system should be the center of mass of molecule BC , atom A lies in the (y,z) plane and moves with velocity u_0 in z direction. Initial position and momentum of A relative to BC are given by the following vectors:

$$\vec{R}^{\circ} \equiv \left\{ 0, b, -(R^{\circ 2} - b^2)^{1/2} \right\},$$

$$\vec{P}_R^{\circ} \equiv \left\{ 0, 0, \mu_{A,BC} u_0 \right\}.$$

(32)

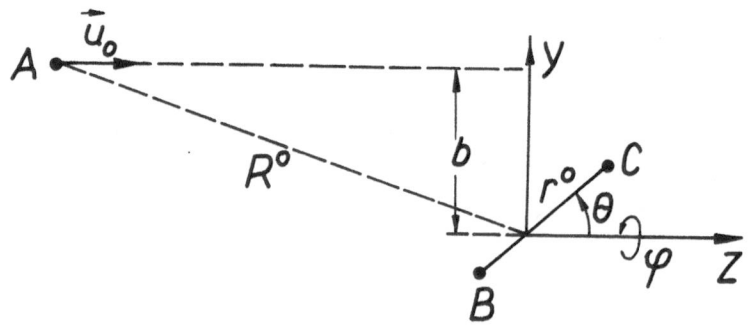

Fig. 3.-6

The remaining 6 initial conditions have to define the initial arrangement of molecule BC by the values of r, θ, φ as well as the molecular momentum by its absolute value and direction.

According to quantum mechanics, disregarding electronic motion the internal state of a molecule is specified by the quantum numbers of vibration and rotation; in classical mechanics the bound-state motion is not quantized. However, it is possible to introduce an element of quantum-mechanical description into the classical treatment by restricting the internal molecular energy to the quantum-mechanically allowed vibration-rotational levels v, J. This approach, being otherwise a purely classical one, is commonly called the quasiclassical method.

Altogether, the initial state of the system is characterized by the following 9 data:

E_{tr} relative translational energy
$= \mu_{A,BC}\, u_o^2 /2$
(CM collision energy)

ν vibrational quantum number of BC

J rotational quantum number of BC

$\left. \right\}$ relevant controllable parameters   (33)

R^o initial distance $A-BC$

b impact parameter

r^o elongation of BC distance
($\triangleq$ vibrational phase)   (34a)

θ^o, φ^o angular position of BC

η^o direction of total molecular momentum at maximal or minimal elongation of BC distance;

the 3 remaining initial data are contained in the special choice of the coordinate system:

$$q_4^o = 0 \quad , \quad p_4^o = p_5^o = 0 .\qquad (34b)$$

Only the 3 parameters E_{tr}, ν and J are "physically relevant", i.e. controllable(at least on principle) in a molecular-beam experiment.

According to the different character of the two groups of parameters, (33) and (34), in the calculations they are handled in different ways:

(1) fixing of E_{tr} , ν , J ;

(2) choice of the remaining parameters by chance ("Monte-Carlo method").

After carrying out a large number of calculations with randomly chosen parameters R°, b , r°, θ°, φ°, η° an averaging procedure for the computed data is added, simulating in this way an idealized molecular-beam experiment.

For each set of initial conditions the canonical equations (31a-c) have to be solved numerically; for this task, efficient computers are necessary. In dependence on the type of system treated, a distance Q_0 is chosen so that atoms being apart from each other by a distance larger than Q_0 do not interact. The integration of the equations of motion is carried out until at least two distances, e.g. R_{AC} and R_{BC} , are larger than Q_0 . When this condition is fulfilled the computation is terminated and the computer signals, in the example mentioned, the formation of a molecule AB . In case that all three distances are larger than Q_0 , the process is counted as dissociative.

Further details on the practice of quasiclassical calculations can be found in the literature /14/.

Analysis and Graphical Representation of Results of
Quasiclassical Calculations

The primary result of the calculations are the coordinates and momenta as functions of time:

$$q_1(t), \ldots, q_6(t), p_1(t), \ldots, p_6(t).$$

For illustration of the classical dynamics of collisions between an atom A and a diatomic molecule BC the following representations can be used:

(1) The unique relationship between the coordinates $\overline{q}_j$, eq. (26), and q_j , eq. (28), enables us to calculate from q_j the $\overline{q}_j(t)$ functions – the <u>trajectories of the particles</u> A, B and C in space.

For the special case of a planar motion (A , B and C lie permanently in a fixed plane) the trajectories can be represented graphically in a direct way. Fig. 3.-7 shows a central collision $(b = 0)$ in the system $H + H_2$. Initially, the molecule carries out an unperturbed zero-point vibration $(v = 0, J = 0)$; when the collision partners approach the H_2 $(\cong BC)$ bond widens and, finally, a new molecule H_2 $(\cong AB)$ is formed moving backwards (with respect to the incoming atom) in the negative z direction. The product molecule is vibrating stronger than the reactant, i.e. translational energy has been partly converted into vibrational energy.

(2) From the coordinates $q_j(t)$ the <u>interparticle distances</u> can be calculated and represented graphically as functions of time; we consider again $H + H_2$.

Fig. 3.-8 shows such diagram for a non-reactive, inelastic process. As a result of the collision, the vibrational energy of molecule BC has become smaller; this energy difference is converted partly to translational, partly to rotational energy.

In the reactive collision shown in fig. 3.-9, a considerable part of the translational energy is transferred to vibrational energy, a smaller part to rotational energy of the product molecule. The collision time is of the order 10^{-14} s, much smaller than the rotational period; the process considered is a direct one (see section 2.4.).

(3) The parametric representation $R_i \equiv R_i(R_j, R_k)$ of the nuclear configurations traversed during the collision is called the <u>system trajectory.</u>

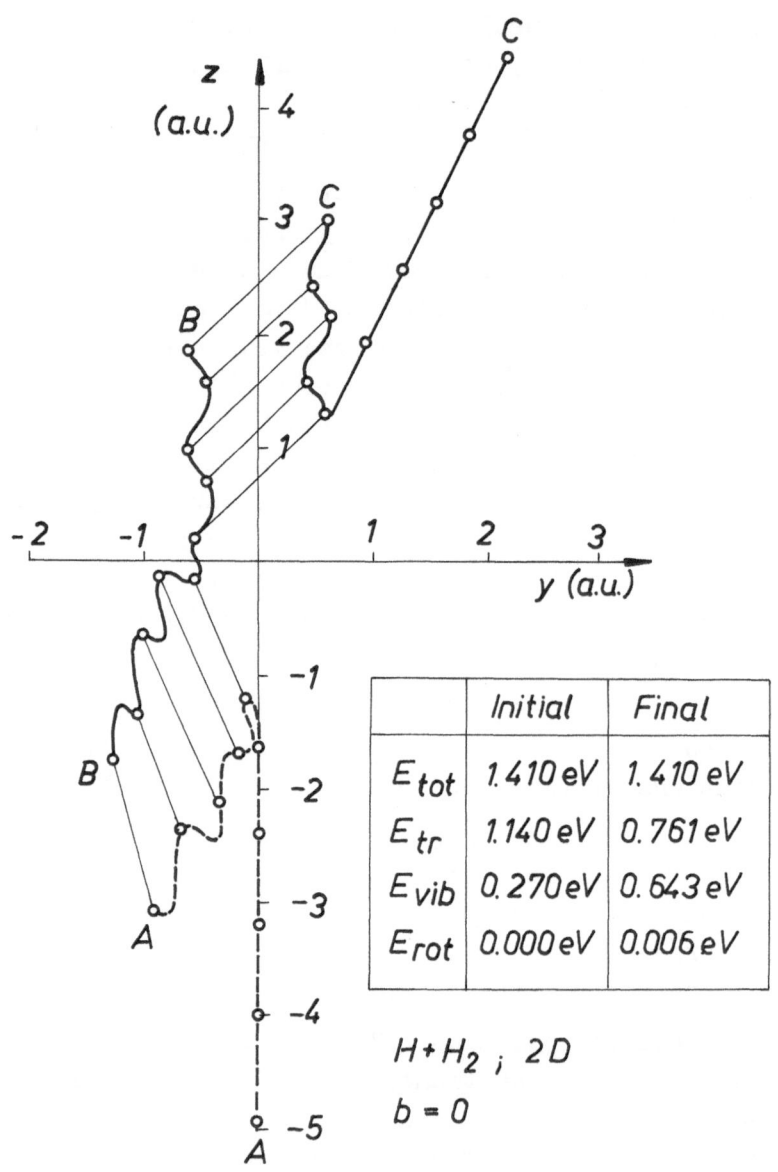

	Initial	Final
E_{tot}	1.410 eV	1.410 eV
E_{tr}	1.140 eV	0.761 eV
E_{vib}	0.270 eV	0.643 eV
E_{rot}	0.000 eV	0.006 eV

$H + H_2$; 2D

$b = 0$

Fig. 3.-7

53

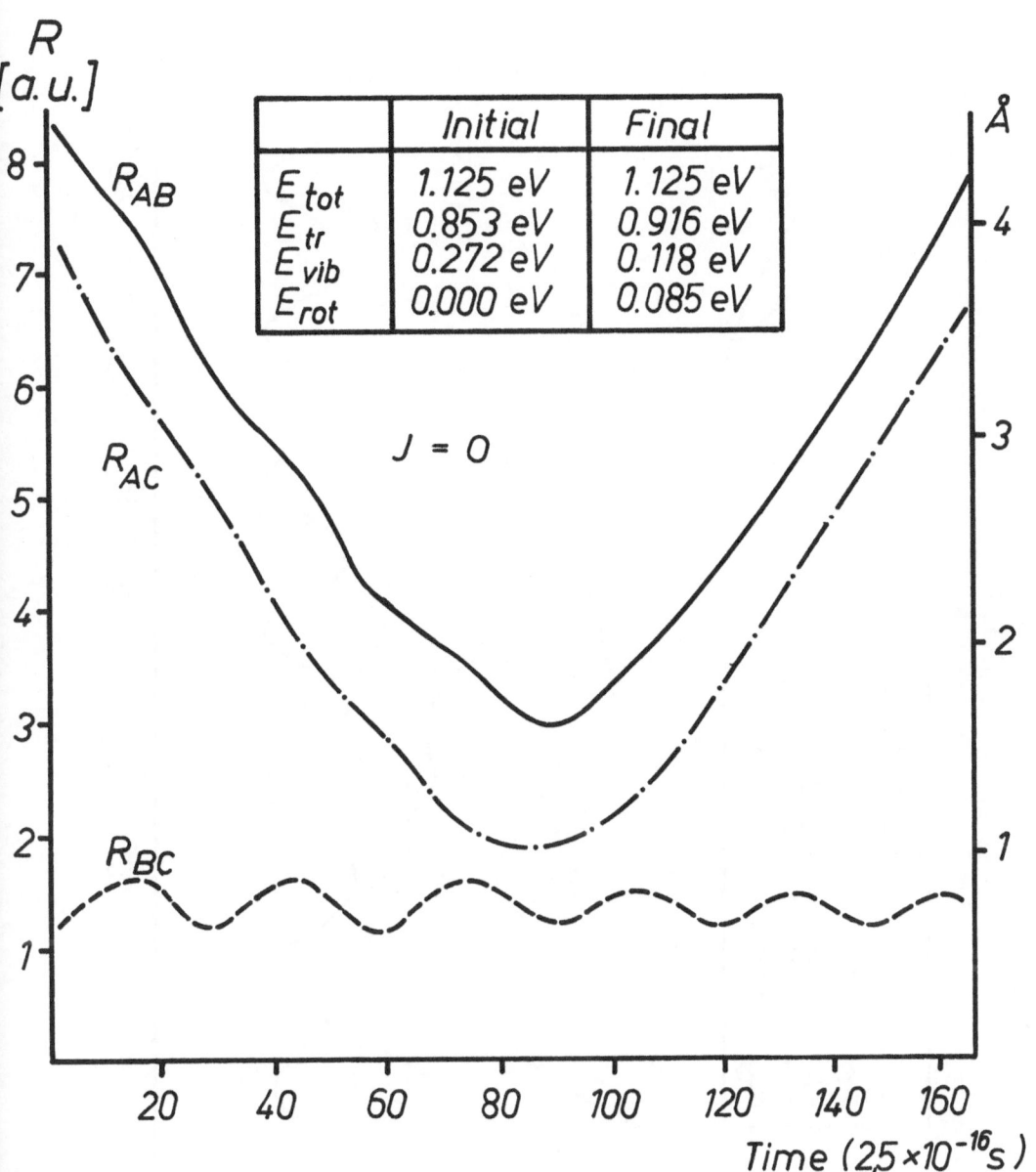

Fig. 3.-8

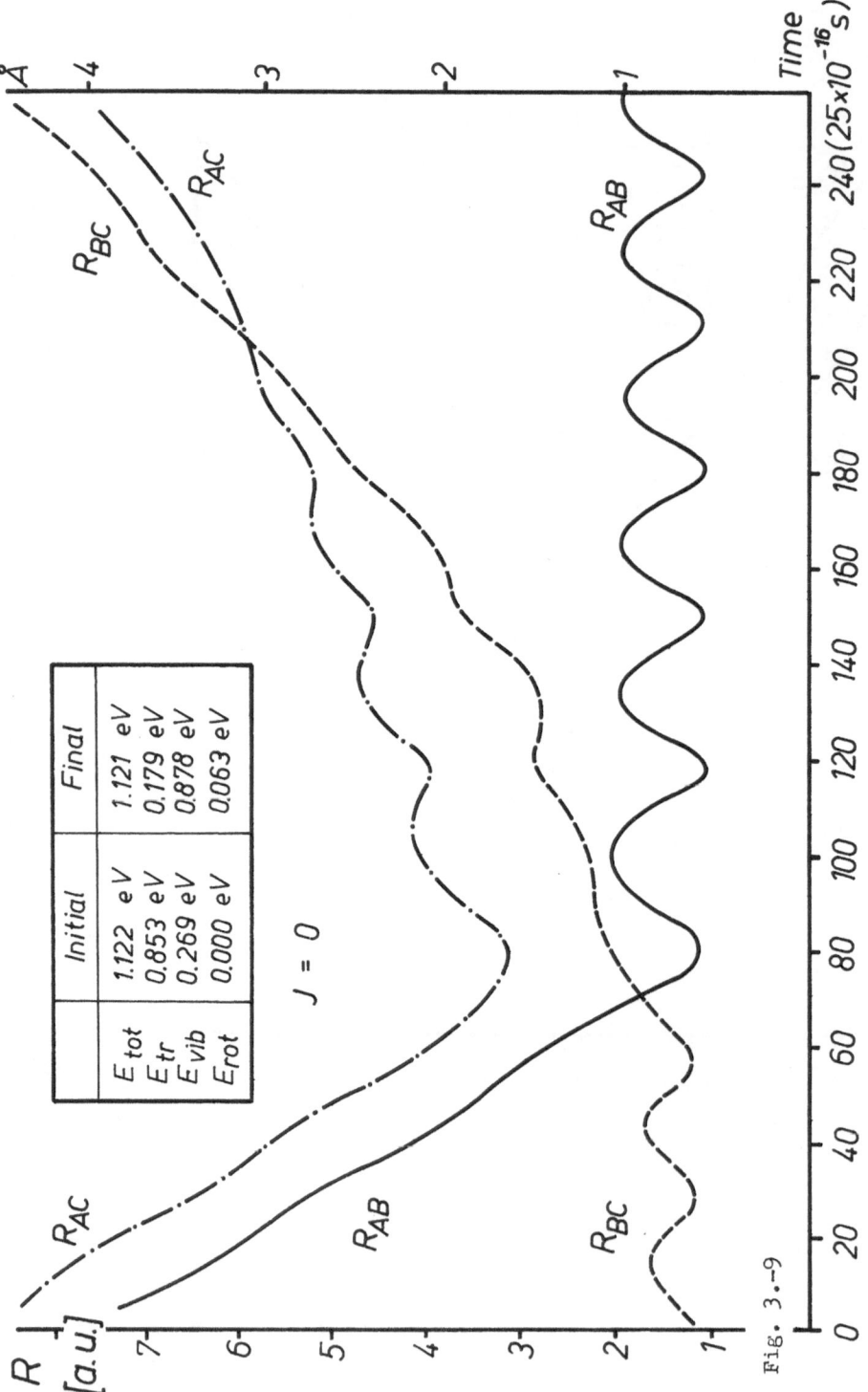

	Initial	Final
E_{tot}	1.122 eV	1.121 eV
E_{tr}	0.853 eV	0.179 eV
E_{vib}	0.269 eV	0.878 eV
E_{rot}	0.000 eV	0.063 eV

$J = 0$

Fig. 3.-9

Fig. 3.-10 corresponds to a collinear motion (all three
particles permanently lie on a fixed straight line)
showing the curve $R_{BC} = R_{BC}(R_{AB})$ in the (R_{AB}, R_{BC})
plane. As can be seen, the process proceeds directly and
translational energy is converted partly to vibrational
energy.

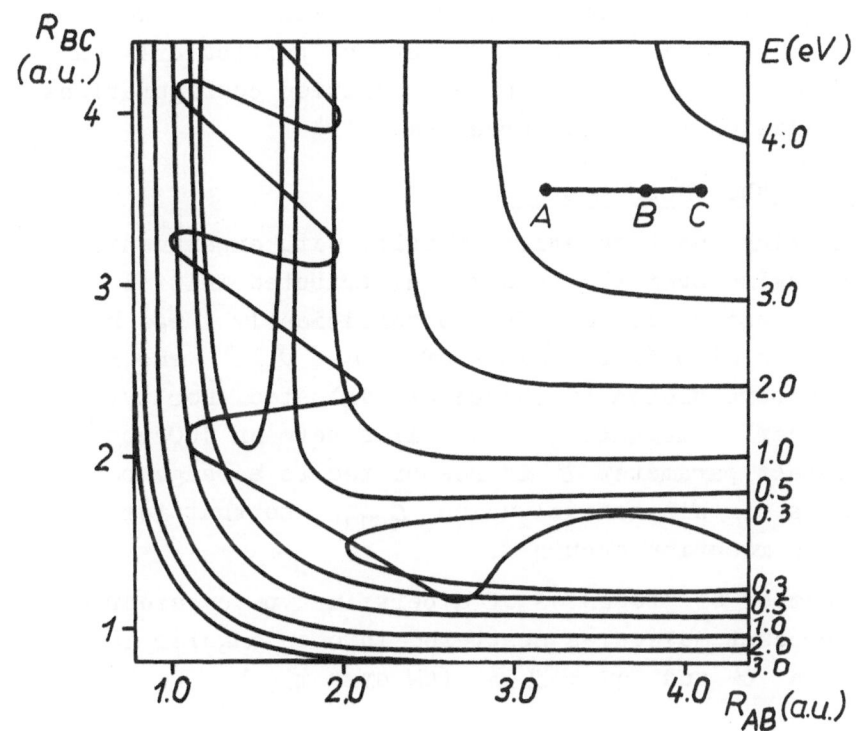

Fig. 3.-10

By the values of the momenta p_i at the instant of termination of the calculation, the different energy portions (translational, vibrational and rotational)[1] as well as the scattering angle are determined. Having carried out this analysis, one frequently converts the final vibrational and rotational energies into corresponding (in general, non-integer) "quantum numbers" using appropriate models (say, Morse oscillator and rigid rotator); by rounding off these values to integers one obtains classical final vibrational and rotational quantum numbers v' and J', respectively.

For each single collision process the calculation provides in this way a lot of important information, particularly on energy-exchange processes, favoured transition configurations of atoms, collision time and mechanism.

Calculation of the Cross Section

For connecting the theoretical results with experimental data, an averaging over the randomly distributed initial conditions is necessary. To this end, a sufficiently large number $\mathcal{N}$ of processes with fixed values of E_{tr}, v, J and remaining parameters chosen by chance are evaluated. For typical cases, the order of magnitude of $\mathcal{N}$ lies between 100 and 1000; the impact parameter b is restricted to a range between 0 and a certain maximal value b_{max} so that for $b > b_{max}$ no reaction occurs[2].

Let the number of processes of a certain type considered, leading to product molecules scattered into an angular interval between ϑ and $\vartheta + \Delta \vartheta$ (CM system), be

---

[1] It should be mentioned that, whereas the translational and internal energy parts are strictly separable from each other as soon as the interaction between the products vanishes, the separation of the internal energy into vibrational and rotational portions can be achieved only by introducing approximations.

[2] This b_{max} can be determined by using model considerations or by test calculations.

$$\Delta \mathcal{N}(E_{tr}, \nu, J; \vartheta) \; ; \tag{35}$$

this number is proportional to the incident particle-beam density,

$$\dot{j}_0 = \frac{\mathcal{N}(E_{tr}, \nu, J)}{\pi \, b_{max}^2} \, , \tag{36}$$

and to the solid-angle interval corresponding to (ϑ, $\vartheta + \Delta \vartheta$),

$$\Delta \Omega = 2\pi \sin \vartheta \, \Delta \vartheta \, . \tag{37}$$

The proportionality factor is called differential cross section:

$$\sigma(E_{tr}, \nu, J; \vartheta)$$

$$= \pi \, b_{max}^2 \, \frac{\Delta \mathcal{N}(E_{tr}, \nu, J; \vartheta)}{\mathcal{N}(E_{tr}, \nu, J) \, 2\pi \sin \vartheta \, \Delta \vartheta} \tag{38}$$

(compare chapter 1.). By integration (or summation, respectively) over all ϑ intervals the total cross section results:

$$\sigma^{total}(E_{tr}, \nu, J)$$

$$= \int \sigma(E_{tr}, \nu, J; \vartheta) \, 2\pi \sin \vartheta \, d\vartheta \, . \tag{39}$$

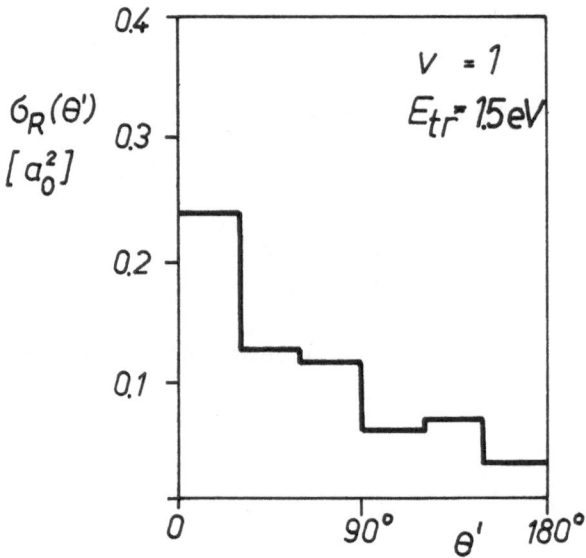

Fig.3.-11

    As an example for a differential reaction cross section calculated by the quasiclassical method we show in fig. 3.-11 a result for the process /15/

$$He + H_2^+ (\nu = 1, J = 0) \longrightarrow HeH^+ + H \qquad (40)$$

counting HeH^+ in arbitrary internal states; the angular interval used is 30°. The form of the cross section corresponds to a stripping mechanism (see section 2.4.).

3.1.4. Examples of Results of Trajectory Calculations

For an illustration of recent advances in quasiclassical trajectory calculations we give a brief account of some interesting results without going into details.

(1) Investigation of the Exchange Reaction $D + H_2 \rightarrow HD + H$

For the differential cross section of this key reaction of chemical kinetics reliable experimental results are available from molecular-beam measurements by GEDDES et al. /16/; no selection of velocity and internal states of reactants and products, however, could be accomplished.

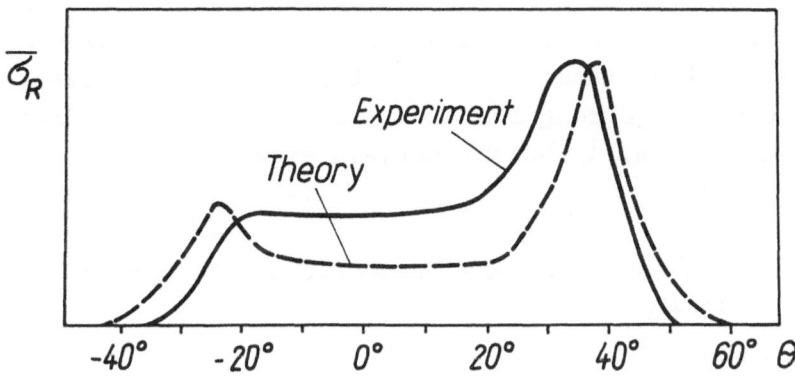

Fig. 3.-12

In fig. 3.-12 we show the experimental curve (L system) in comparison with the theoretical one determined by averaging of the calculated data for specific reactant velocities and internal states using appropriate distribution functions /17/; θ denotes the laboratory scattering angle. The agreement is pretty good.

(2) Studies of Reactive and Dissociative Processes
$He + H_2^+$

The system $He + H_2^+$ represents a simple model for
the study of elementary processes in ion-molecule reactions.
Experimental information exists on the energetics(endo-
ergicity 0.8 eV for ground-state reactants and products),
cross sections and other characteristics /18/.

Recently, extensive quasiclassical calculations have been
carried out providing a detailed understanding of the
dynamics of the elementary processes /19/. As an example
we compare in fig. 3.-13 the experimental /18a/ and theoretical /19a/ total cross sections for exchange reaction and
dissociation as functions of the collision energy E_{tr} for
reactants in the third excited vibrational state. Fig. 3.-14
shows the experimental /18b/ and theoretical /19/ differential reactive cross section averaged over the different
vibrational states contained in a vibrationally non-selected
reactant beam. The strong forward peak indicates a dominant
stripping mechanism. Agreement between experiment and theory
is good likewise.

(3) Calculations of Elementary Processes in the System
$F + H_2, HD, D_2$

The elementary processes

$$F + H_2(v, J) \longrightarrow FH(v', J') + H , \qquad (41a)$$

$$F + HD(v, J) \Big\langle \begin{array}{l} FH(v', J') + D, \\ FD(v', J') + H, \end{array} \qquad (41b)$$

$$F + D_2(v, J) \longrightarrow FD(v', J') + D \qquad (41c)$$

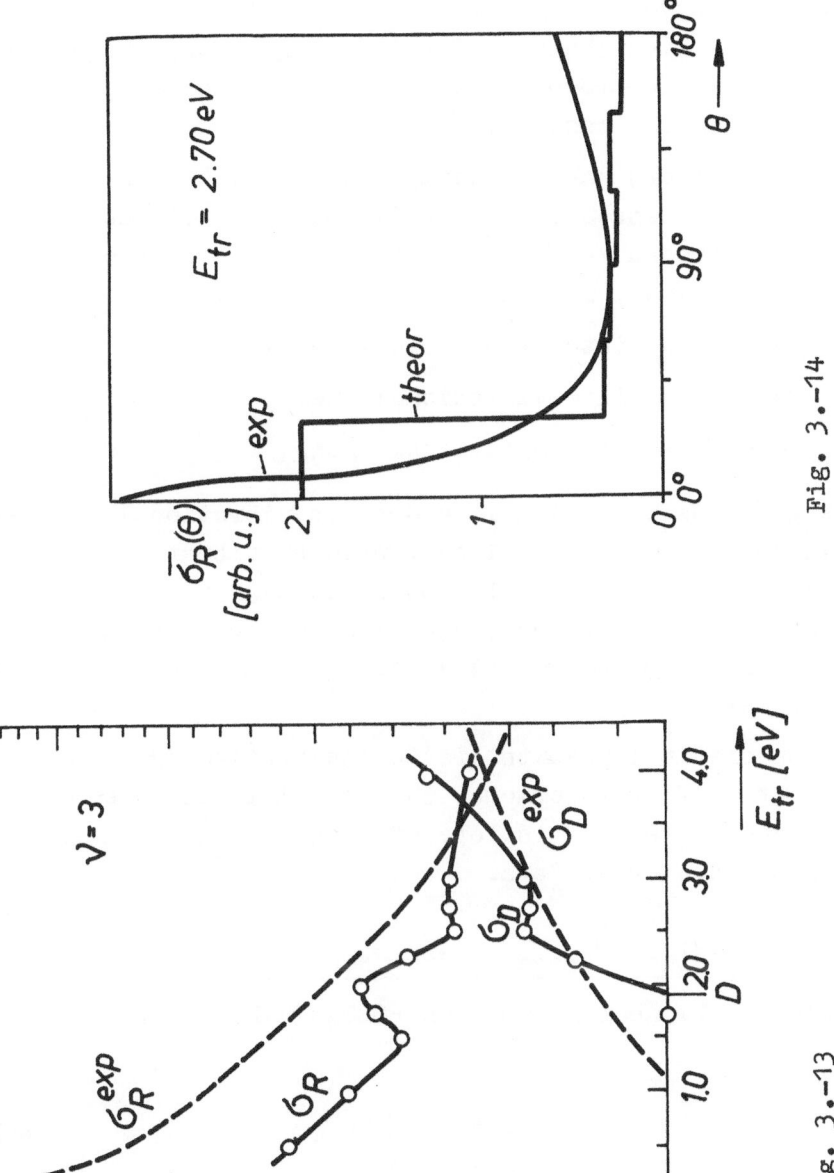

Fig. 3.-14

Fig. 3.-13

are exoergic by about 1.3 eV for ground-state reactants and
products; the distribution of this energy among the degrees
of freedom of the products is of considerable practical
interest (chemical lasers).

Several authors have investigated the processes (41) and
the reverse processes theoretically by means of the quasiclassical approach /20/. The available energy, according to
these calculations, is converted as follows:

11 - 18 % into translational energy,

77 - 80 % into vibrational energy,

4 - 10 % into rotational energy

of the products resulting in a drastic vibrational popula-
tion inversion: while the lowest product vibrational levels
$(\nu' = 0,1)$ are almost empty, the higher levels
$(\nu' = 2,3,4)$ are strongly occupied. One can use this
property of the processes (41) for pumping of chemical
lasers /21a/.

The theoretical results are in qualitative agreement with
experiment /21/; furthermore, the calculations have provided
predictions concerning isotope effects for the processes
(41b) /20a/ and other data.

(4) Theory of the Hydrogen-Iodine Reaction

Since the pioneering work of BODENSTEIN, the reaction

$$H_2 + I_2 \rightarrow 2HI \qquad (42)$$

was considered as the classical example of an elementary
bimolecular reaction. In 1967, however, SULLIVAN /22/ has
shown this reaction to be a two-step process:

$$I_2 + M \longrightarrow 2I + M \, , \qquad\qquad (43a)$$

$$H_2 + 2I \left\langle \begin{array}{l} \nearrow \, 2\,HI \qquad\qquad\qquad (43b) \\ \searrow \, H_2I + I \longrightarrow 2\,HI \, . \quad (43c) \end{array} \right.$$

The reason for this behaviour of the system $H_2 + I_2$ cannot be understood within the framework of purely energetic (static) considerations but only from dynamical investigations.

Indeed, quasiclassical trajectory calculations /23/ lead to the conclusion that process (42) is "dynamically forbidden": by reason of the structure of the potential energy surface the system trajectory almost never reaches the product valley even for high collision energies. The corresponding rate constant is too low by 5 orders of magnitude. On the other hand, the reaction steps (43) are much more "easy" to carry through leading to a rate constant of the correct order of magnitude. Furthermore, the calculations show that some of the trajectories pass in fact through an intermediate complex H_2I .

This example clearly indicates the importance of dynamical studies in clearing up the mechanism of chemical reactions.

3.2. Elements of Quantum–Mechanical Methods

At present, the quasiclassical method explained in
section 3.1. enables, for a given potential–energy surface,
the calculation of adiabatic elementary processes in three-
and four–atomic systems without substantial difficulties; it
leads to important insights into the collision dynamics and
provides numerical results semi–quantitatively. The quasiclassical approach suffers, however, from principal defects,
since it does not take into account quantum phenomena like
tunneling, resonance and interference; strictly speaking,
in investigations of molecular collision processes quantum
mechanics must be applied. The error in the calculation of
macroscopic quantities (rate constants) caused by neglect
of collisional quantum effects cannot be estimated quantitatively nowadays.

Because of the great difficulties in solving the quantum–
mechanical scattering problem, at present almost exclusively
simple models (e.g., the collinear model) and comparatively
rough approximations are practicable. Many of the methods
are rather complicated; for this reason, we treat in the
following only some basic concepts and simple approximations
referring frequently to the literature /5c,9,13/.

3.2.1. Correspondence of Classical and Quantum–Mechanical Theories

The quasiclassical method describes the dynamics of a
collision process by the system trajectory – i.e., a curve
in the configuration space of the nuclei representing the
successive nuclear arrangements passed through by the system
during the interaction process. Formally, the problem corresponds to the motion of a mass point in a multi–dimensional
space under the influence of a potential depending on the
internal variables.

To introduce the corresponding quantum-mechanical description, we consider for reasons of simplicity at first the free motion ($V \equiv 0$) of a particle of mass m along a straight line chosen as the x axis of our coordinate system[1]. Because of the Heisenberg uncertainty relation,

$$\Delta x \cdot \Delta p \geqslant \hbar/2 , \tag{1}$$

in contrast to classical mechanics the particle cannot have at the same time a definite coordinate x and a definite momentum p - consequently, to the particle cannot be ascribed a trajectory in the classical sense. Let the particle have the definite momentum p - i.e., in the absence of forces, the energy $E = p^2/2m$ - then its motion can be described by the wavefunction

$$\psi_p (x,t) = exp\left\{ (\pm ipx - iEt)/\hbar \right\} \tag{2}$$

as solution of the time-dependent Schrödinger equation,

$$-\frac{\hbar^2}{2m} \frac{\partial^2}{\partial x^2} \psi_p(x,t) = i\hbar \frac{\partial}{\partial t} \psi_p(x,t) . \tag{3}$$

The probability density of finding the particle at the point x depends neither on x nor on time:

---

[1] All considerations can be extended easily to the motion in three-dimensional space; the following general results remain valid.

$$w(x,t) \equiv \psi_p^* \, \psi_p = 1 \,, \qquad (4)$$

i.e., the particle cannot be localized in space (maximal uncertainty of coordinate). Consequently, the plane wave (2) is not immediately suitable for describing the motion of a particle; for this purpose, a wavefunction is necessary allowing at every instant t a localization within a certain limited region of space. Since arbitrary linear combinations of plane waves (2) are solutions of the Schrödinger equation (3) likewise, one can try to obtain such localized probability distributions $w(x,t)$ by superpositon of plane waves using appropriate amplitude factors $g(p)$ /5c,9a/:

$$\Psi(x,t) = h^{-\frac{1}{2}} \int_{-\infty}^{\infty} g(p) \, \psi_p(x,t) \, dp \,; \qquad (5)$$

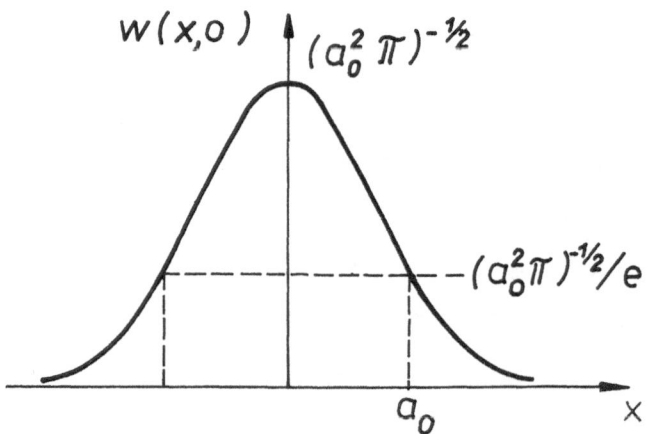

Fig. 3.-15

wavefunctions of this type (in the form of Fourier integrals) are called "wavepackets". We assume the probability density $w(x,t)$ at time $t = 0$ to have the form of a normalized Gaussian distribution around a point chosen as origin of the coordinate system (fig. 3.-15):

$$w(x,0) \equiv \Psi^*(x,0)\,\Psi(x,0) = (a_o^2 \pi)^{-\frac{1}{2}} \exp\{-(x/a_o)^2\}\,;\,(6)$$

the parameter a_o is a measure of the uncertainty Δx of the particle localization. By the distribution $w(x,0)$ the wavefunction $\Psi(x,0)$ is determined with the exception of a phase factor; writing

$$\Psi(x,0) = h^{-\frac{1}{2}} \int_{-\infty}^{\infty} g(p)\, \exp(ipx/\hbar)\, dp$$

$$= (a_o^2 \pi)^{-\frac{1}{4}} \exp\{-(x^2/2a_o^2) + (ip_o x/\hbar)\}$$

(7)

leads to a current density $j_o \equiv Re\{\Psi^*[\hbar/im]\,\partial\Psi/\partial x\}$ $= w(x,0)\,p_o/m$. The phase factor $\exp(ip_o x/\hbar)$ takes care of the correspondence between the wavepacket and the moving particle: the maximum of the wavepacket runs with the velocity $u_o = p_o/m$ in the positive x direction.

Having settled in this way the initial wavefunction $\Psi(x,0)$, the amplitude $g(p)$ can be determined by the reverse Fourier transformation

$$g(p) = h^{-\frac{1}{2}} \int_{-\infty}^{\infty} \Psi(x,0)\, \exp(-ipx/\hbar)\, dx$$

$$= (a_o^2/\pi\hbar^2)^{\frac{1}{4}} \exp\{-(p-p_o)^2 a_o^2/2\hbar^2\}\,;$$

(8)

this function, likewise, has the form of a Gaussian with its maximum at the mean momentum p_o . Analogously to the localization uncertainty, the quantity $\hbar/a_o$ can be considered as a measure of the momentum uncertainty Δp – the width of the distribution $|g(p)|^2$; in accordance with relation (1) we have here $\Delta x \cdot \Delta p \approx \hbar$. The more the particle is localized ($a_o = \Delta x$ small), the broader is the range Δp of momenta to be included in the wavepacket describing its motion; corresponding to this momentum uncertainty, the mean value (expectation value) of the energy, calculated using the wavefunction (7), is given by

$$\overline{E} = \langle \Psi | -\frac{\hbar^2}{2m} \frac{\partial^2}{\partial x^2} | \Psi \rangle = \frac{p_o^2}{2m} + \frac{\hbar^2}{2ma_o^2} \ . \quad (9)$$

Since Ψ contains waves ψ_p with $p > p_o$ as well as such with $p < p_o$, as a necessity the distribution $w(x,t)$ broadens in the course of time. Substituting the amplitude function (8) in the expression (5), integrating and taking the square modulus gives :

$$w(x,t) = (a^2 \pi)^{-\frac{1}{2}} \exp \left\{ -(x - u_o t)^2 / a^2 \right\} \quad (10)$$

with

$$a = a_o \left[1 + (\hbar t / m a_o^2)^2 \right]^{\frac{1}{2}} . \quad (10a)$$

The center of the distribution $w(x,t)$ moves with constant velocity u_o in the positive x direction decreasing in height and increasing in width. This "smearing" of the wavepacket proceeds the faster the smaller the mass m of the particle is.

The free motion of a sufficiently localized wavepacket corresponds, as we have seen, closely to the motion of a free mass point in classical mechanics: quantum-mechanical probability distribution and classical mass point both run on a straight line with constant velocity according to Newton's law. This correspondence also holds for a motion under the influence of a potential V . Confining ourselves again to one dimension, let us consider the expectation value of the coordinate X :

$$\bar{x}(t) = \int\limits_{-\infty}^{\infty} \Psi^*(x,t)\, \Psi(x,t)\, x\, dx \ . \tag{11}$$

For the force-free case, $V \equiv 0$, using a distribution of the form (10) initially localized at point X_o , the relation $\bar{x}(t) = x_o + u_o t$ results. For $V \neq 0$ the wavefunction $\Psi(x,t)$ is determined by the Schrödinger equation

$$\hat{H}\, \Psi(x,t) = i\hbar\, \frac{\partial}{\partial t}\, \Psi(x,t) \tag{12}$$

with the Hamiltonian

$$\hat{H} = -\frac{\hbar^2}{2m}\, \frac{\partial^2}{\partial x^2} + V(x)\, , \tag{12a}$$

instead of the Schrödinger equation (3). Differentiating eq. (11) twice, using the Schrödinger equation (12), integrating by parts and taking into consideration the fact that $\Psi(x,t)$ vanishes for $x \longrightarrow \pm \infty$, one arrives at the Ehrenfest equation

$$m \frac{d^2}{dt^2} \overline{x}(t) = \overline{F}(t) \qquad (13)$$

where

$$\overline{F}(t) \equiv \int_{-\infty}^{\infty} |\Psi(x,t)|^2 \left(-\frac{\partial V}{\partial x}\right) dx \qquad (13a)$$

denotes the average force. This relation shows complete analogy to classical mechanics - but generally the quantum-mechanical "path" of the particle is different from the classical one because of $\overline{F}(t) \neq -(\partial V/\partial x)_{x=\overline{x}}$. If, however, the localization is sufficiently sharp and the potential varies slowly over distances $\Delta x = x - \overline{x}$ of the order of a_0 so that the condition

$$\frac{1}{2} \left| \frac{\partial^3 V}{\partial x^3} \right|_{x=\overline{x}} \overline{(\Delta x)^2} \ll \left| \frac{\partial V}{\partial x} \right|_{x=\overline{x}} \qquad (14)$$

is fulfilled, equation (13) can be written in the following approximate form:

$$m \frac{d^2}{dt^2} \overline{x}(t) \approx -\frac{\partial V(\overline{x})}{\partial \overline{x}} \; ; \qquad (15)$$

the "center" $\overline{x}$ of the wavepacket moves then nearly along the classical path. Consequently, the inequality (14) represents a condition for the applicability of classical mechanics (see section 3.1.1.).

3.2.2. Time-Dependent Scattering Theory

By reason of the hermiticity of the Hamiltonian the
normalization integral $\langle \Psi | \Psi \rangle$ of a wavefunction remains
constant in the course of time; consequently, the motion
from one moment t_o to another moment t can be described
by a transformation with a unitary operator $\hat{U}$ /5c,9/:

$$\Psi(x,t) = \hat{U}(t,t_o)\,\Psi(x,t_o) \tag{16}$$

with

$$\hat{U}^+(t,t_o)\,\hat{U}(t,t_o) = \hat{U}(t,t_o)\,\hat{U}^+(t,t_o) = 1 . \tag{17}$$

This time-evolution operator (dynamical operator) satisfies
the integral equation

$$\hat{U}(t,t_o) = 1 - \frac{i}{\hbar}\int_{t_o}^{t}\hat{H}\,\hat{U}(t',t_o)\,dt' \tag{18}$$

which includes the initial condition

$$\hat{U}(t_o,t_o) = 1 . \tag{19}$$

For solving the time-dependent Schrödinger equation (12),
at first the initial state, e.g. in the form (7), at time
t_o has to be settled, localized in a region of constant
potential (no forces). Then, the evolution operator trans-
forms the wavefunction from one time instant to another.

In fig. 3.-16 we show a schematic picture of a scattering process in a system $A + BC$ considered as one-
dimensional (along the "reaction coordinate" or minimum-
energy path). The wavepacket moves towards the potential
barrier and there it will be partly reflected and partly

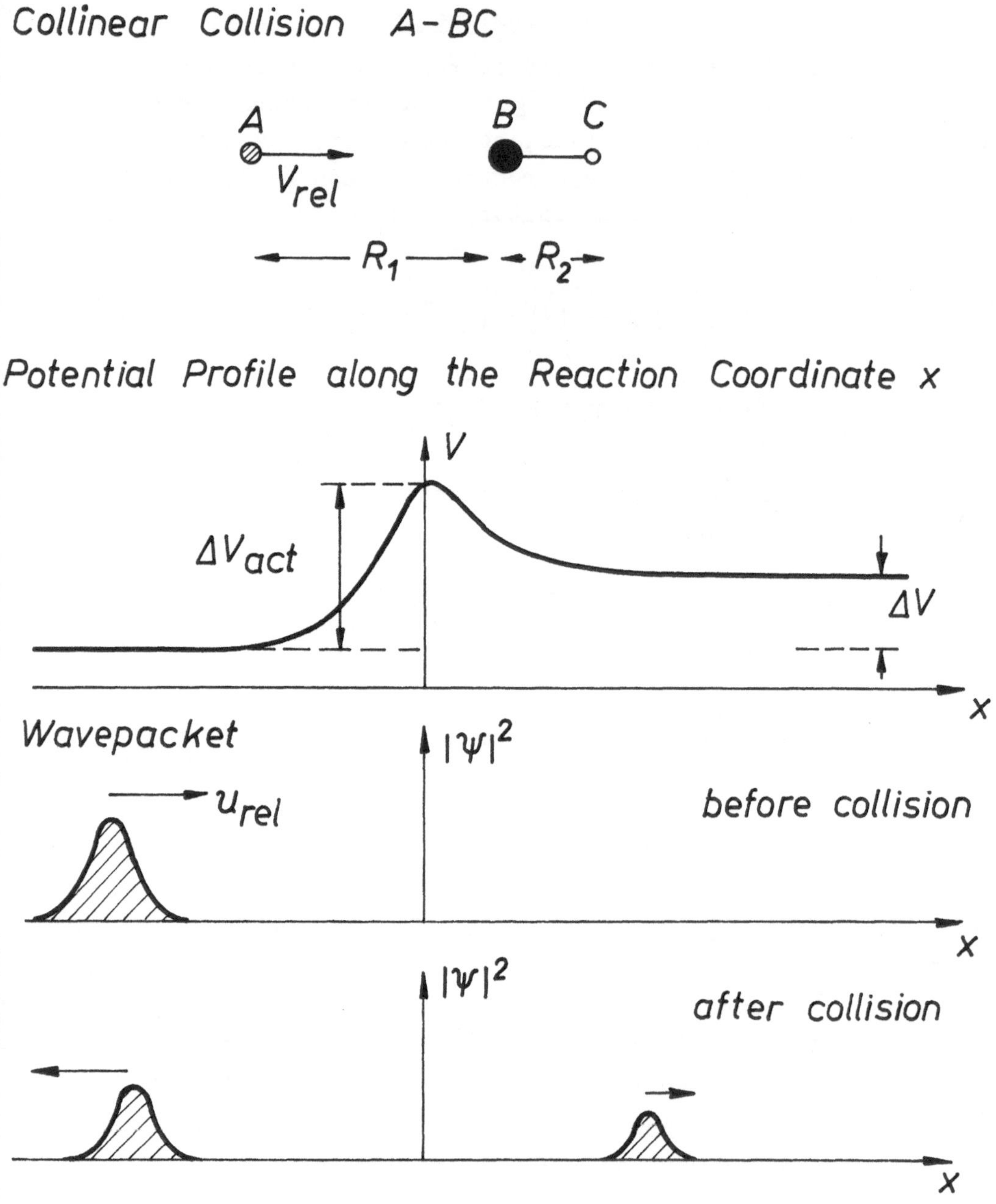

Collinear Collision A-BC
A
B
C
V_{rel}
R_1
R_2
Potential Profile along the Reaction Coordinate x
V
ΔV_{act}
ΔV
x
Wavepacket
$|\psi|^2$
u_{rel}
before collision
x
$|\psi|^2$
after collision
x

Fig. 3.-16

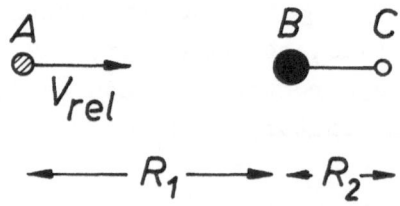

transmitted. After a sufficiently long time, for $t = t_e$, a reflected and a transmitted wavepacket move in the outer regions of constant interaction potential. The so-called scattering operator (S matrix) /9,13/[1],

$$\hat{S} \equiv \hat{U}(t_e, t_o), \qquad (20)$$

contains all information about the scattering process.

Treatment of Collinear Collisions $A + BC$

Using the notation of section 3.1.3. (eqs. (27,28,30)), the Hamiltonian for collinear configurations of the system $A + BC$ after separation of the center-of-mass motion is given by

$$\hat{H} = -\frac{\hbar^2}{2\mu_{BC}} \frac{\partial^2}{\partial r^2} - \frac{\hbar^2}{2\mu_{A,BC}} \frac{\partial^2}{\partial R^2} + V(r, R) ; \quad (21)$$

r is the internuclear distance of the molecule BC and R the distance of atom A to the center of mass of BC . For large values of R the potential V becomes identical with the diatomic interaction potential $V_{mol}(r)$, and the wavefunction of the initial state can be written as the product of a wavepacket $F(R)$ for the relative motion $A - BC$ and a vibrational eigenfunction $\eta_v(r)$ of the free molecule BC ,

$$\Psi_v(r, R; t_o) = F(R) \eta_v(r) , \qquad (22)$$

----

[1] Strictly speaking, the matrix elements of the S operator are defined by passing to the limits $t_o \to -\infty$, $t_e \to +\infty$ in the matrix elements of $\hat{U}(t_e, t_o)$.

$\eta_\nu(r)$ being a solution of the vibrational Schrödinger equation

$$\left\{-\frac{\hbar^2}{2\mu_{BC}}\frac{d^2}{dr^2} + V_{mol}(r)\right\}\eta_\nu(r) = e_\nu\,\eta_\nu(r). \quad (23)$$

The wavepacket $F(R)$ is to be formed according to eq. (7), taking R instead of X , and localized at a sufficiently large distance R_o . Since the time-dependance of the wavefunction Ψ_ν usually cannot be found in analytical form, it is necessary to solve the Schrödinger equation

$$\hat{H}\,\Psi_\nu(r,R;t) = i\hbar\frac{\partial}{\partial t}\,\Psi_\nu(r,R;t) \quad (24)$$

with the initial condition (22) by means of numerical methods /24/.

To give an idea of the motion of such a wavepacket we show some results /24c/ for the hydrogen exchange reaction $H + H_2 \longrightarrow H_2 + H$. For a clear demonstration, the probability densities $|\Psi_\nu|^2$ in the reactant valley and in the product valley have been integrated over the coordinates R_{BC} and R_{AB} , respectively, obtaining in this way "one-dimensional probability densities". Figs. 3.-17 and 3.-18 represent some selected pictures – "snapshots" of the wavepacket's motion[1].

---

[1] It should be noted that the expression $\Delta R_{AB}\int|\Psi_\nu|^2 dR_{BC}$ in the reactant valley gives the probability for finding the atoms A and B at a distance between R_{AB} and $R_{AB} + \Delta R_{AB}$; the corresponding holds in the product valley.

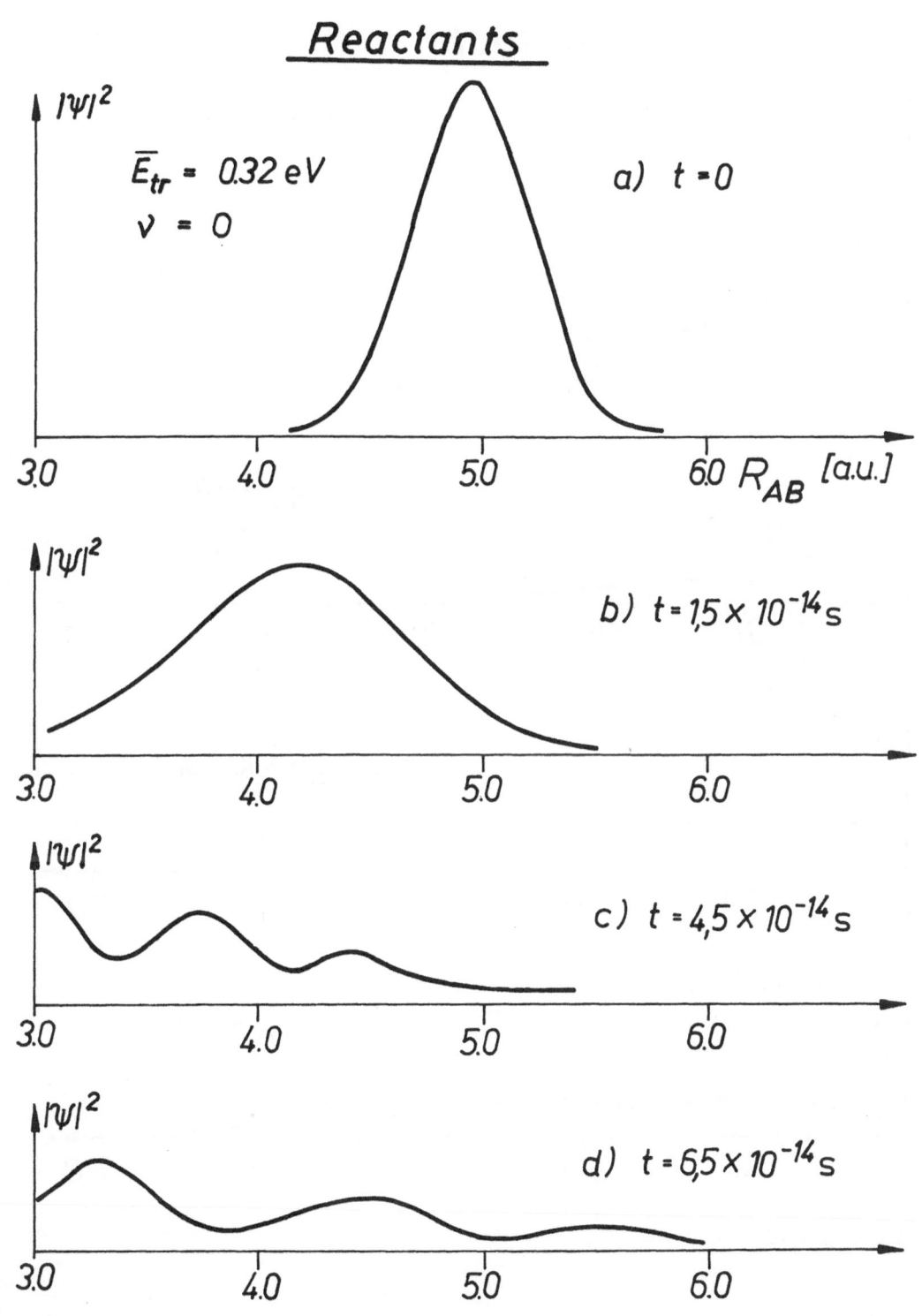

Fig. 3.–17

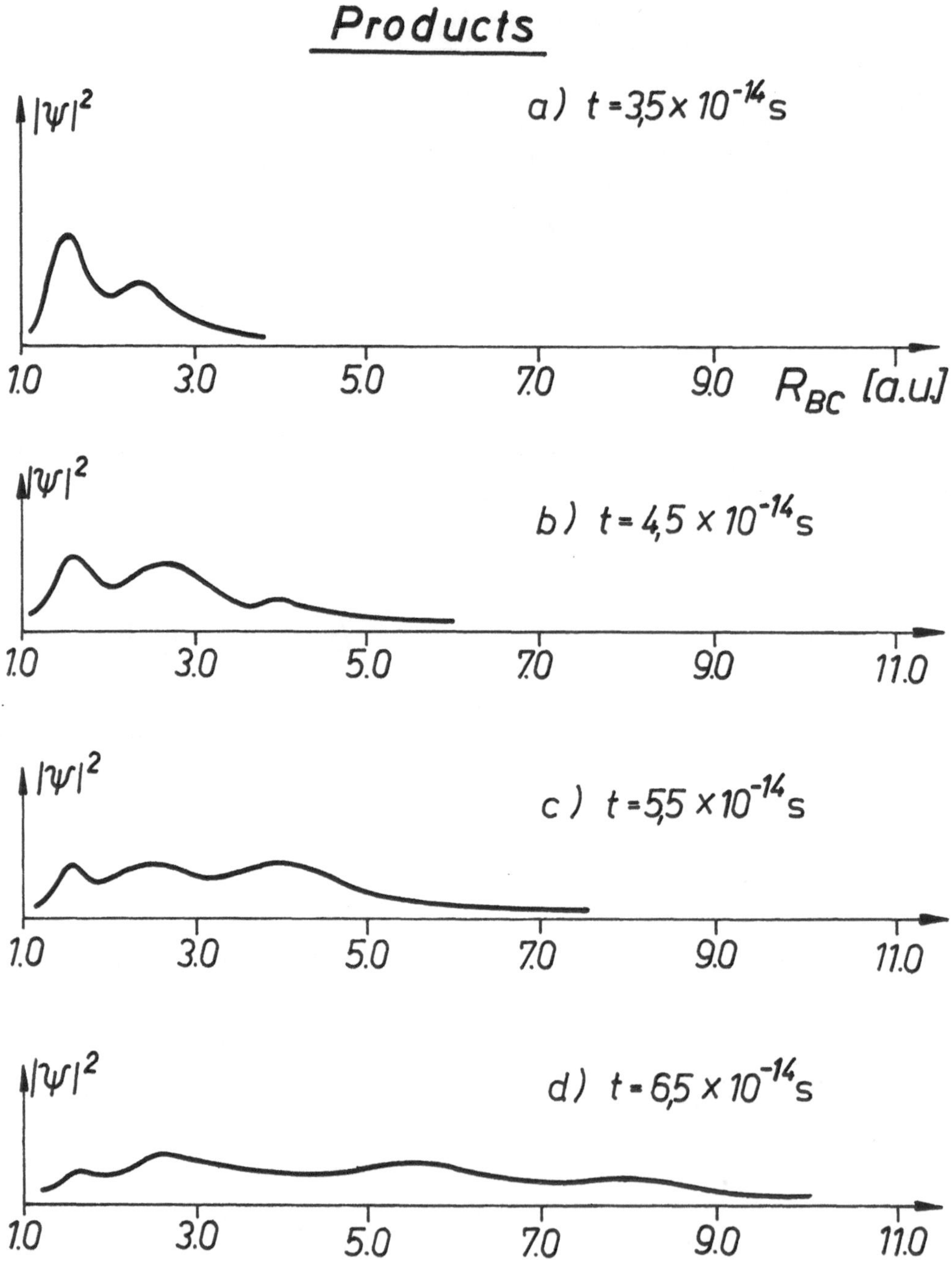

Products
$|\psi|^2$
a) t = 3,5 × 10⁻¹⁴ s
1.0
3.0
5.0
7.0
9.0
R_{BC} [a.u]
$|\psi|^2$
b) t = 4,5 × 10⁻¹⁴ s
1.0
3.0
5.0
7.0
9.0
11.0
$|\psi|^2$
c) t = 5,5 × 10⁻¹⁴ s
1.0
3.0
5.0
7.0
9.0
11.0
$|\psi|^2$
d) t = 6,5 × 10⁻¹⁴ s
1.0
3.0
5.0
7.0
9.0
11.0

Fig. 3.-18

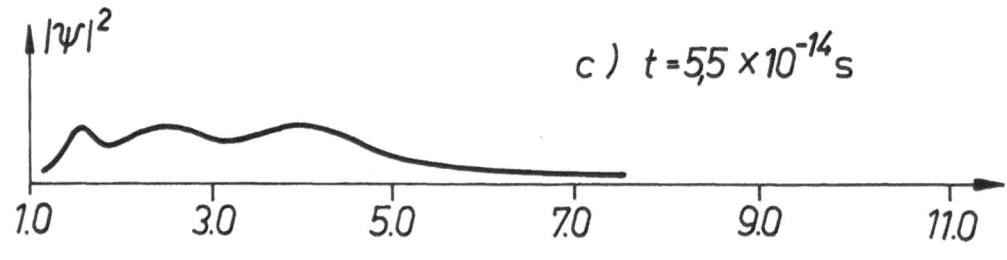

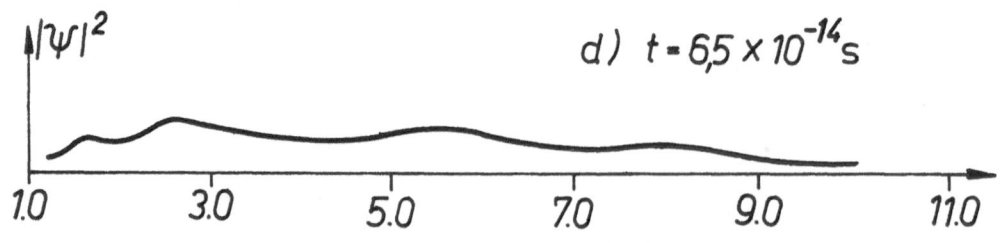

In fig.3.-17a is shown a Gaussian initial wavepacket centered at $R_{AB} \approx 5$ a.u. in the reactant valley. The motion proceeds towards the interaction region (i.e., to the vicinity of the saddle point), the quick broadening can be seen in fig.3.-17b. Interference between incoming and reflected parts of the wavepacket leads to the wavy structure of the distribution in figs. 3.-17c,3.-17d,where the process is almost finished: the reflected portions of the wavepacket representing elastically and inelastically scattered re-actants move in the direction of increasing distance R_{AB}.

The probability distribution in the interaction region and in the product valley can be seen from figs. 18. In fig. 3.-18a already a considerable part of the wavepacket has entered the region of strong interaction (R_{AB} and R_{BC} small); the transmitted part begins to run into the product valley, it is quickly broadening and becoming smeared over a large area(figs.3.-18b-d). Integration of this distribution in the product valley over R_{BC} from 0 to ∞ directly gives the reaction probability.

3.2.3. Stationary Scattering Theory

By reason of the immediate analogy to classical mechanics, the time-dependent scattering theory sketched above has the advantage of being physically obvious and interpretable. Its disadvantages are the complications in solving the scattering problem arising from the time-dependence and mainly the fact that, because of the superposition character of the wavefunction, only averaged quantities (over a certain momentum range) can be calculated - some more subtle quantum effects (resonance phenomena, e.g.) are not obtained.

Therefore, one can try to consider scattering states with sharp values of energy and momentum - as limiting cases of spatially very extended wavepackets. Such formula-tion of the problem leads to the stationary Schrödinger equation. Again we treat at first the one-dimensional case; the subsequent extension to three-dimensional problems (including reactions) can be given here only in a rather formal way disregarding some mathematical questions as well as details of numerical procedures. Finally, we mention a few examples of the application to chemical elementary processes and compare with quasiclassical results.

3.2.3.1. <u>One-Dimensional Scattering</u>

We consider a wavepacket superposed of momentum eigenfunctions of a very small p interval $(p'-\varepsilon, p'+\varepsilon)$, only in this interval the amplitude function $g_\varepsilon(p)$ is non-vanishing. The particle described by this wavepacket has an almost well-defined momentum p', because of the uncertainty relation $\Delta x \cdot \Delta p \geqslant \hbar/2$ it extends in coordinate space over a large region. In the limiting case $\varepsilon \to 0$ and $g_\varepsilon(p) \longrightarrow \delta(p-p')$ the wavepacket degenerates to an infinitely extended stationary scattering wave with well-defined momentum:

$$\Psi(x,t) \longrightarrow \psi_{p'}(x)\,\exp(-ip'^2 t/2m\hbar). \quad (25)$$

The wavefunction $\psi_{p'}(x)$ satisfies the time-independent Schrödinger equation,

$$\hat{H}\,\psi_{p'}(x) = E'\,\psi_{p'}(x), \quad (26)$$

with $E' = p'^2/2m$ and certain boundary conditions discussed below. The solutions of eq. (26) are not normalizable in the usual sense; for the free motion, e.g., we have

$$\langle \psi_{p'} | \psi_{p''} \rangle = \delta(p'-p'') . \qquad (27)$$

Therefore, we consider the stationary solutions (25) as auxiliary quantities from which actual states can be obtained by superposition with appropriate amplitude factors $g(p)$. The functions (25) themselves may be set in relation to a homogeneous monoenergetic particle beam of sufficiently low density. It can be shown /13a/ that from the solutions of eq. (26) the same information is obtained as from a time-dependent treatment by means of a wavepacket with a very small but finite momentum interval.

For the sake of simplicity we assume it to have a potential $V(x)$ of finite range:

$$V(x) \xrightarrow[|x| \to \infty]{} c \, |x|^{-\lambda} \qquad (\lambda > 1) \quad (28)$$

(c = const). The boundary conditions to be imposed upon the solutions of eq. (26) concern the asymptotic behaviour. Let the incoming particles move in positive x direction towards the scattering center $(V \neq 0)$; in the stationary description this situation corresponds to a constant particle beam entering from $x = -\infty$ represented by a plane wave. Under the action of the potential this beam splits into a transmitted (i.e., in x direction continuing) and a reflected part. Thus, for the asymptotic form of the stationary scattering wave we take

$$\psi_p(x) \simeq \begin{cases} \exp(ipx/\hbar) + \varrho_p \exp(-ipx/\hbar) & \\ \qquad\qquad\qquad\qquad x \to -\infty & \\ \tau_p \exp(ipx/\hbar) & x \to +\infty . \end{cases} \qquad (29)$$

The (generally complex-valued) reflection and transmission coefficients, ϱ_p and τ_p, respectively, are connected via the probability conservation condition:

$$|\varrho_p|^2 + |\tau_p|^2 = 1 , \tag{30}$$

i.e., the number of scattered (reflected or transmitted) particles must be equal to the total number of particles.

Scattering by a Square-Well Barrier. Tunnel Effect

As a model for one-dimensional collision processes we consider the scattering of a particle by a square-well potential barrier (/5c,9/ and other textbooks):

$$V(x) = \begin{cases} 0 & \text{in region I } (x<0) \\ V_a = const & \text{in region II } (0 \leqslant x \leqslant a) \\ 0 & \text{in region III } (x>a) \end{cases} \tag{31}$$

with $V_a > 0$ (see fig. 3.-19).

At first, the Schrödinger equation (26) is solved in the three regions I – III of constant potential: the solutions in I and III are plane waves $exp(\pm ipx/\hbar)$, in II one also obtains exponential functions $exp(\pm \beta_0 x)$ with $\beta_0 = [2m(V_a - E)/\hbar^2]^{1/2}$. Therefore, the general solution in the regions I – III is:

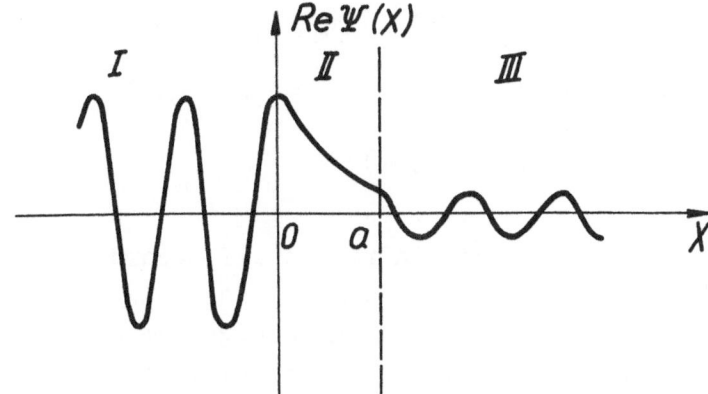

Fig. 3.-19

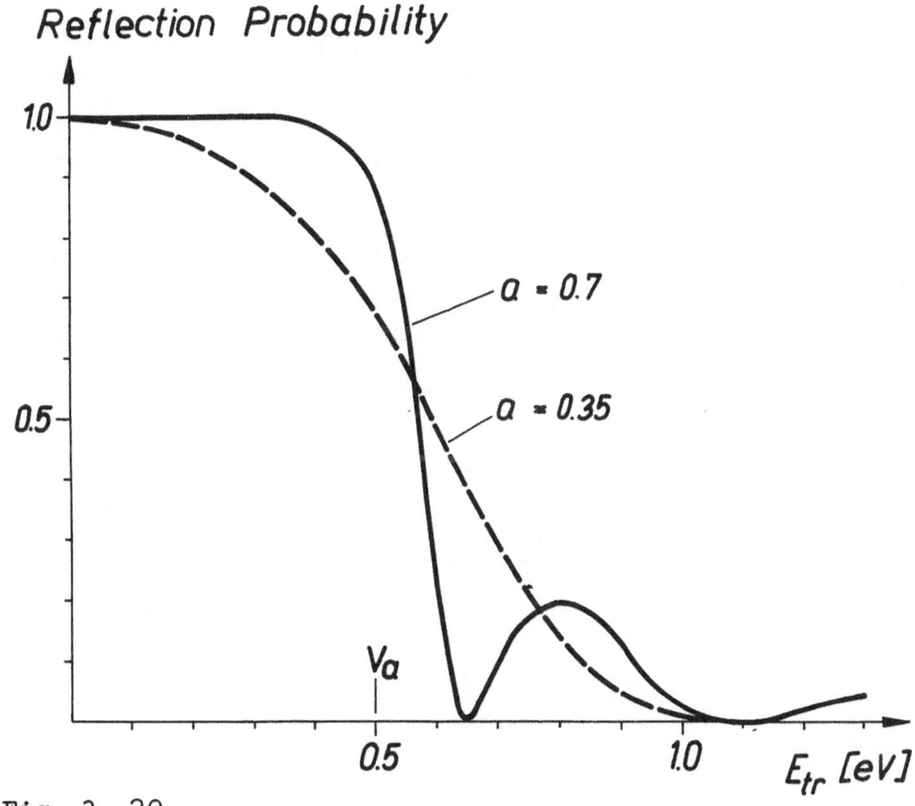

Reflection Probability

Fig. 3.-20

$$\psi^{I}(x) = \exp(ipx/\hbar) + \varrho_p \exp(-ipx/\hbar),$$

$$\psi^{II}(x) = A_o \exp(\beta_o x) + B_o \exp(-\beta_o x), \qquad (32)$$

$$\psi^{III}(x) = \tau_p \exp(ipx/\hbar),$$

with only an outgoing wave in region III according to the boundary condition (29). The coefficients $\varrho_p, \tau_p, A_o, B_o$, being generally complex-valued and energy-dependent, are determined by the conditions of continuity of the wave-function and its first derivative at the separating points of the three regions, i.e., at $x = 0$ and $x = a$:

$$\psi^{I}(0) = \psi^{II}(0), \qquad \psi^{II}(a) = \psi^{III}(a),$$

$$\frac{d\psi^{I}}{dx}\bigg|_{x=0} = \frac{d\psi^{II}}{dx}\bigg|_{x=0}, \quad \frac{d\psi^{II}}{dx}\bigg|_{x=a} = \frac{d\psi^{III}}{dx}\bigg|_{x=a}. \qquad (33)$$

From this system of linear equations the four coefficients can be evaluated easily; we give here the solutions for the reflection and transmission probabilities, $R_p \equiv |\varrho_p|^2$ and $T_p \equiv |\tau_p|^2$:

$$R_p = \begin{cases} \sinh^2|\beta_o|a / (\sinh^2|\beta_o|a + \varkappa_o^2) & E < V_a, \\ \sin^2|\beta_o|a / (\sin^2|\beta_o|a + \varkappa_o^2) & E > V_a, \end{cases} \qquad (34a)$$

$$T_p = 1 - R_p, \qquad (34b)$$

with $\varkappa_o^2 \equiv 4\,E\,|E-V_a|/V_a^2$

For an illustration of this result we show in fig. 3.-20 the reflection probability for a proton scattered by barriers of height $V_a = 0.5$ eV having widths $a = 0.35$ a.u. and $a = 0.7$ a.u. In dependence on the energy E, pronounced quantum effects are observed: the reflection probability is smaller than unity (correspondingly, the transmission probability larger than zero) even for energies below the barrier height ("tunnel effect"), and also for higher energies the curves differ from the classical ones ($R_p^{class} = 1$ for $E < V_a$, $R_p^{class} = 0$ for $E > V_a$). For $a = 0.7$ a.u. interference effects are evident.

Fig. 3.-19 demonstrates qualitatively the form of the real part of the wavefunction for $E < V_a$. In regions I and III we have undamped (free) waves; in region II the amplitude decreases but remains different from zero because of the finite width of the barrier (tunneling).

3.2.3.2.   Three-Dimensional Elastic Scattering

Having discussed some features of stationary scattering theory for the simple one-dimensional case, we now admit more degrees of freedom. The next step is the treatment of the elastic scattering of two particles, A and B, with an interaction potential $V(R)$ (see section 3.1.2.).

The Schrödinger equation in relative coordinates after separation of the center-of-mass motion,

$$\hat{H}\,\psi(\vec{R}) \equiv \left\{-\frac{\hbar^2}{2\mu}\,\Delta_R + V(R)\right\}\psi(\vec{R}) = E\,\psi(\vec{R}), \quad (35)$$

is to be solved with appropriate boundary conditions. In analogy to one-dimensional scattering we imagine a constant particle (A) beam, described by a plane wave $exp(i\vec{p}\,\vec{R}/\hbar)$

impinging upon the scattering center (B). For large distances
from the scattering center the wave function should be a
superposition of the incoming wave and a (by reason of
symmetry) spherically scattered wave outgoing from B :

$$\psi(\vec{R}) \underset{R \to \infty}{\simeq} \exp(i\vec{p}\,\vec{R}/\hbar) + f(\vartheta)\,\frac{1}{R}\,\exp(ipR/\hbar). \quad (36)$$

The square modulus of the "scattering amplitude" $f(\vartheta)$
gives the probability for scattering of A by the angle ϑ ,
i.e. the differential scattering cross section

$$\sigma(\vartheta) = \left| f(\vartheta) \right|^2 \qquad (37)$$

(the ratio of scattered and incoming particle-current
densities).

In practice, the calculation of the wavefunction $\psi(\vec{R})$
with asymptotic form (36) is accomplished, to any desired
degree of accuracy, by expansion into Legendre polynomials
(so-called partial-wave expansion) /5c,9,12,13/.

For a general, formally simple formulation of the scattering problem defined by eqs. (35) and (36) it is useful
to convert the Schrödinger equation into an integral equation which contains automatically the condition (36). This
reformulation (see /9,13/) leads to the Lippmann-Schwinger
equation,

$$\psi(\vec{R}) = \exp(i\vec{p}\,\vec{R}/\hbar) + \int d^3\vec{R}'\, G_o(\vec{R},\vec{R}';E)\,\psi(\vec{R}')\,V(R'); \quad (38)$$

the integral kernel $G_o(\vec{R},\vec{R}';E)$ – the Green's function –
is a solution of the differential equation

$$\left(-\frac{\hbar^2}{2\mu}\Delta_R + E\right) G_o(\vec{R},\vec{R}';E) = \delta(\vec{R}-\vec{R}') \qquad (39)$$

and can, if one imposes the condition of an outgoing-spherical-wave behaviour, be given explicitly in our case as

$$G_o(\vec{R},\vec{R}';E) = -\frac{2\mu}{4\pi\hbar^2}\frac{1}{|\vec{R}-\vec{R}'|}\exp(ip|\vec{R}-\vec{R}'|/\hbar). \qquad (40)$$

Even more concise, one may write the Lippmann-Schwinger equation (38) in an operator form:

$$\psi(\vec{R}) = \exp(i\vec{p}\vec{R}/\hbar) + \hat{G}_o(E)V(R)\psi(\vec{R}) \qquad (38')$$

using the Green's operator

$$\hat{G}_o(E) \equiv \left(E + \frac{\hbar^2}{2\mu}\Delta_R + i\varepsilon\right)^{-1} \qquad (40')$$

the coordinate representation of which is the Green's function (40). In application of eq. (38'), after operating and solving one has to pass to the limit $\varepsilon \to 0$ (ε real positive).

3.2.3.3. Rearrangement Scattering (Reactions)

Now, we consider a system consisting of two colliding composite particles (e.g., two molecules); according to chapter 1. and section 2.3., elastic, inelastic and reactive (rearrangement) scattering is possible:

$$P(i) + Q(j) \xrightleftharpoons{} \begin{array}{l} P(i) + Q(j) \\ P(i') + Q(j') \\ X(l) + Y(m) \, , \end{array} \qquad (41)$$

disregarding here dissociation. By the index α we denote
(see section 2.3.) the entrance channel, i.e. the partitioning $\{ P(i), Q(j) \}$ of the total system including the
internal states, by β one of the possible exit channels.
Where a more precise characterization is necessary, to the
channel index a collective index a, b ,... for the quantum
numbers of the internal states of the particles in channels
α, β ,... will be added.

The separated reactants and products in the center-of-
mass coordinate system are described by the Hamiltonians
$\hat{H}_\alpha$ and $\hat{H}_\beta$, respectively, which in each case are
composed of an operator $\hat{T}_R$ for the relative kinetic
energy and an operator $\hat{H}^\circ$ determining the internal motions. Hence, for a discussion of the asymptotic behaviour
of the wavefunction a channel-dependent partitioning of $\hat{H}$
is useful:

$$\hat{H} = \hat{H}_\alpha + V_\alpha = \hat{H}_\beta + V_\beta \qquad (42)$$

with

$$\hat{H}_\alpha = \hat{T}_{R_\alpha} + \hat{H}_\alpha^\circ \, , \quad \hat{H}_\beta = \hat{T}_{R_\beta} + \hat{H}_\beta^\circ \, ; \qquad (42a)$$

the distances of the centers of mass of P and Q and of X and Y are denoted by R_α and R_β, respectively. We suppose the potentials V_α and V_β to vanish for large distances faster than R_α^{-1} and R_β^{-1}, respectively.

For the internal states of reactants and products the following Schrödinger equations are to be solved:

$$\hat{H}_\alpha^o \chi_{\alpha,a} = E_{int}^{\alpha,a} \chi_{\alpha,a} \ , \qquad (43a)$$

$$\hat{H}_\beta^o \chi_{\beta,b} = E_{int}^{\beta,b} \chi_{\beta,b} \ ; \qquad (43b)$$

the total energy E is obtained as the sum of such eigenvalue E_{int} and the relative translational energy:

$$E = E_{int}^{\alpha,a} + E_{tr}^{\alpha,a} = E_{int}^{\beta,b} + E_{tr}^{\beta,b} \ , \qquad (44)$$

$$E_{tr}^{\alpha,a} = p_{\alpha,a}^2 / 2\mu_\alpha \ , \qquad E_{tr}^{\beta,b} = p_{\beta,b}^2 / 2\mu_\beta \ , \qquad (44a)$$

where μ_α and μ_β are the reduced masses of the particle pairs $P-Q$ and $X-Y$, respectively.

Generalizing the consideration on elastic scattering we imagine a stationary current of reactants P impinging on a reactant Q. As long as there is no interaction, the internal states of the reactants P and Q are described by a function $\chi_{\alpha,a}$ and the relative motion $P-Q$ by a plane wave $exp(i\vec{p}_{\alpha,a}\vec{R}_\alpha/\hbar)$; the incoming wave in the entrance channel consequently has the form

$$\Phi_{\alpha,a} = exp(i\vec{p}_{\alpha,a}\vec{R}_\alpha/\hbar)\chi_{\alpha,a} \ . \qquad (45)$$

88

For a finite-range interaction, the complete scattering wave-function at large distances R_α is a superposition of the function (45) and the spherical waves corresponding to all possible elastic and inelastic processes:

$$\Psi_{\alpha,a} \underset{R_\alpha \to \infty}{\sim} \Phi_{\alpha,a}$$

$$+ \sum_{a'} f_{\alpha,a \to \alpha,a'}(\Omega_\alpha)\, \chi_{\alpha,a'}\, R_\alpha^{-1} \exp(ip_{\alpha,a'} R_\alpha/\hbar); \tag{46}$$

in the exit channels β it is a pure sum of spherical waves for the different internal states of the products:

$$\Psi_{\alpha,a} \underset{R_\beta \to \infty}{\sim} \sum_b f_{\alpha,a \to \beta,b}(\Omega_\beta)\, \chi_{\beta,b}\, R_\beta^{-1} \exp(ip_{\beta,b} R_\beta/\hbar) \tag{47}$$

(Ω_α and Ω_β denote the directions $\vartheta_\alpha, \varphi_\alpha$ and $\vartheta_\beta, \varphi_\beta$ respectively, in channels α and β).

Consequently, the problem consists in solving the Schrödinger equation

$$\hat{H}\, \Psi_{\alpha,a} = E\, \Psi_{\alpha,a} \tag{48}$$

with the conditions (46) and (47) for the asymptotic behaviour of $\Psi_{\alpha,a}$. Then, the differential cross section for the process $\alpha,a \to \beta,b$ is determined by the square modulus of the scattering amplitude $f_{\alpha,a \to \beta,b}(\Omega_\beta)$ multiplied by the quotient of the velocities of scattered and incoming particles:

$$\sigma_{\alpha,a \to \beta,b}(\Omega_\beta) = \frac{p_{\beta,b}\, \mu_\alpha}{p_{\alpha,a}\, \mu_\beta} \left| f_{\alpha,a \to \beta,b}(\Omega_\beta) \right|^2 \tag{49}$$

representing the ratio of scattered and incoming fluxes.

Finally, we give the generalization of the Lippmann-Schwinger equation to reactive processes; it reads in operator form:

$$\Psi_{\alpha,a} = \Phi_{\alpha,a} + \hat{G}_\alpha(E) V_\alpha \Psi_{\alpha,a} \tag{50}$$

with the channel-dependent Green's operator

$$\hat{G}_\alpha(E) \equiv (E - \hat{H}_\alpha + i\varepsilon)^{-1}. \tag{50a}$$

The details of the formal scattering theory starting from eq. (50) as well as practical procedures for its solution cannot be discussed here /13/.

3.2.4. Examples of Quantum-Mechanical Calculations of Reactive Elementary Processes

While the quantum-mechanical calculation of elastic scattering processes raises no difficulties and leads to very good agreement with experiment, the treatment of inelastic and reactive collisions is far from being well developed. Because of the mathematical **complexity, rigorous quantum** computations are restricted at present to the simplest systems and models (e.g., the collinear model). In contrast to the classical approach, the hitherto existing quantum results for real processes are not yet as complete as to allow for direct comparison with experiment.

For reactive processes in three-atomic systems of the type $A + BC$, using the collinear model very accurate solutions have been obtained. In fig. 3.-21, the quantum-mechanical probability for the collinear $H + H_2$ exchange reaction, calculated by JOHNSON/25/, is compared with the corresponding

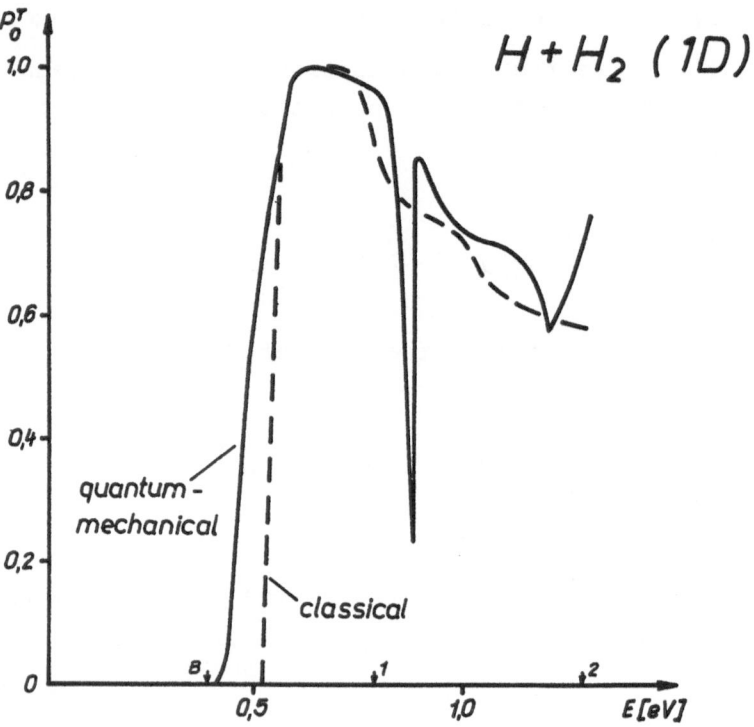

Fig. 3.-21

classical result; P_0^T denotes the total probability for
reactant molecules in the vibrational ground state to give
product molecules in arbitrary vibrational states. Pronounced
quantum effects are observed: (1) The quantum reaction **threshold** is by about 0.1 eV lower than the classical one; for
low energies, therefore, classical mechanics is insufficient
in accord with the considerations in section 3.1.1. (2) **An-**
other striking difference is seen at about 0.9 eV and 1.2 eV
total energy where the quantum-mechanical curve exhibits
"resonances", i.e. sharp local minima. Apart from this, however, the overall agreement of the quantum and classical
reaction probabilities is relatively good.

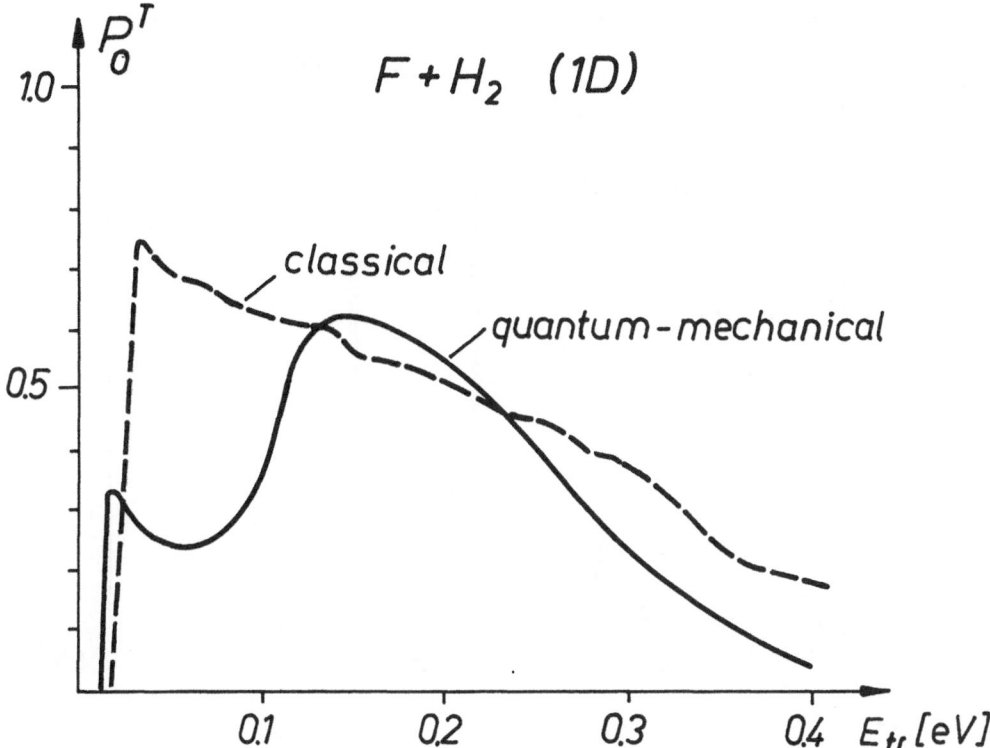

Fig. 3.-22

As a second example of results given by the collinear model we show in fig. 3.-22 corresponding data of SCHATZ et al. /26/ for the $F + H_2$ exchange reaction, eq. (3.1.-41a); again there are marked differences for low energies ($E_{tr} < 0.1$ eV) but also for higher ones ($E_{tr} > 0.25$ eV). Besides, the quantum transition probabilities to specific vibrational states of the product molecule HF, as can be seen from fig. 3.-23, are quite different from the classical data.

Concerning the physical significance of results obtained by using the collinear model, it should be pointed out that on the one hand, because of the elimination of some degrees of freedom, such treatments are appropriate for studying cer-

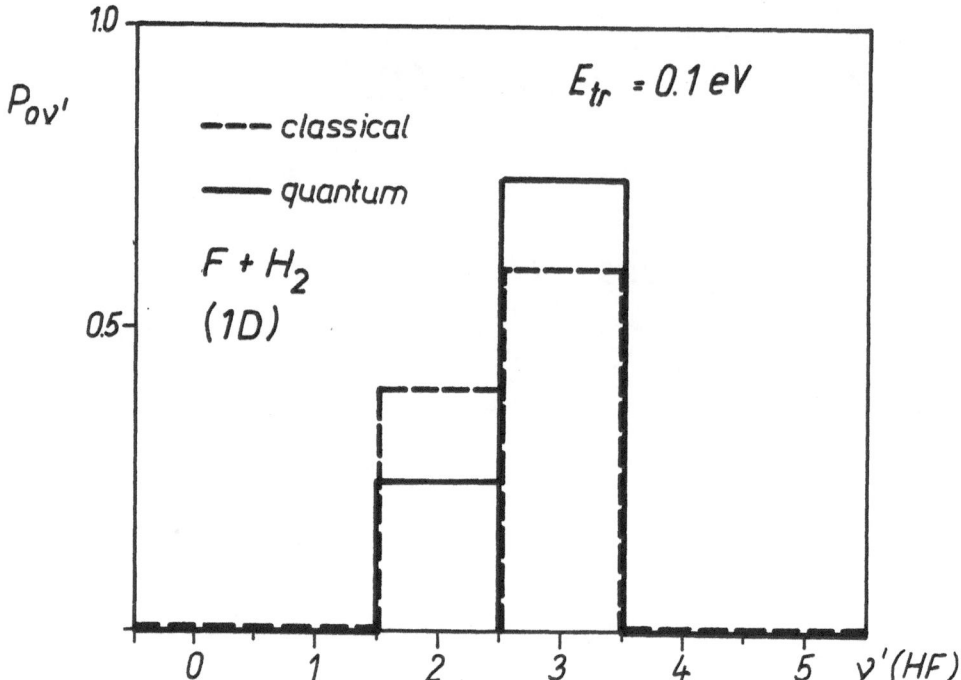

Fig. 3.-23

tain specific quantum effects "purely" — on the other hand, however, the findings are not immediately transferable to the actual three-dimensional problem.

For more than two nuclear degrees of freedom, only very recently the first accurate calculations have been carried through, the test example was again the $H + H_2$ exchange reaction. After some work restricting the motion to coplanar nuclear configurations /27/ and a rough, preliminary computation by WOLKEN and KARPLUS /28a/ for geometrically unrestricted nuclear configurations, more sophisticated treatments by KUPPERMANN and SCHATZ /28b/ and by ELKOWITZ and WYATT /28c/ are now available. Most interestingly, the total reaction cross section obtained (see fig. 3.-24) agrees rather

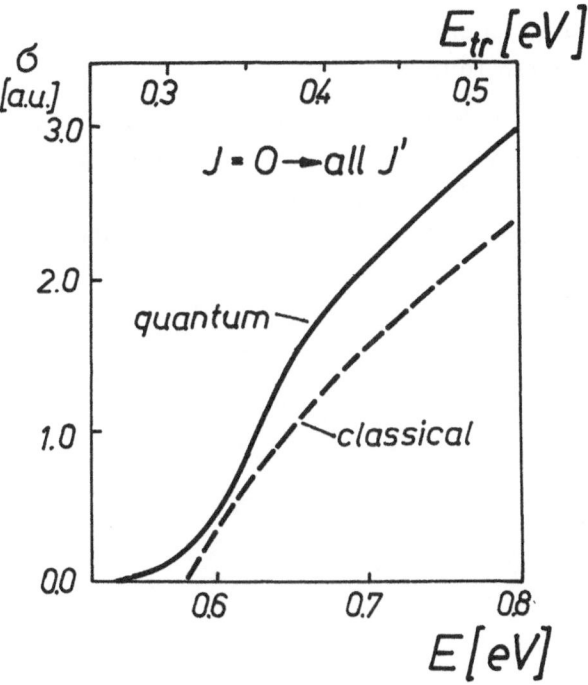

Fig. 3.-24

well with classical trajectory data /14a/; compared with the collinear case, the difference in threshold energies is even smaller. These results are very promising with respect to the reliability and practical utility of classical trajectory calculations.

All other work treating reactive scattering in systems with more than two nuclear degrees of freedom until now is based on comparatively rough approximations. We mention here in particular the so-called distorted-wave Born (DWB) approach which gives apparently, when applied to reactive scattering of molecular systems, merely qualitative results /29/.

In summary, the classical approach seems to reproduce, in the sense of an average, qualitatively the overall shape of quantum cross sections for electronically adiabatic processes; however, resonance phenomena and, in general, also the correct threshold behaviour are not obtained.

In the case of electronically non-adiabatic processes the electronic motion cannot be treated independently of the nuclear motion but must be explicitly taken into account when studying the dynamics. This problem will be discussed in some detail in chapter 5.

4. Classical-Limit and Semiclassical Approaches to the Calculation of Molecular Collisional Transition Probabilities

As has been demonstrated in section 3.1., classical mechanics has proved to be a very useful and, to some extent, reliable tool for the interpretation and prediction of the dynamics of atomic and molecular elementary processes. However, it suffers from serious shortcomings; in particular, it gives usually a wrong low-energy (threshold) behaviour which is crucial for low-temperature rate constants and it does not describe non-adiabatic transitions as well as other specific quantum effects like resonances (compare section 3.2.).

Furthermore, there are other inherent difficulties concerning the quantized nuclear degrees of freedom (vibration, rotation). Trajectory calculations are based on integration of the classical equations of motion using initial conditions which correspond as near, as possible to the quantum states of the reactants. To this end, one may introduce "action-angle variables" $\hbar\tilde{n}$, w [1] /11/ and put the initial values $\tilde{n}_1$ equal to the corresponding quantum numbers n_1 treating the initial angle variables w as being distributed with uniform probability in the classically accessible range[2]. The trajectories calculated for a certain set of initial conditions are then analysed according to the final values $\tilde{n}_2$ and grouped into subsets attributed to definite quantum states n_2; in the simplest procedure, $\tilde{n}_2$ is counted as being equal

---

[1] The symbols n, w denote in the following, as usual, sets of quantum numbers or coordinates, respectively.

[2] Speaking on this choice of initial conditions, the term "quasiclassical approach" is commonly used (section 3.1.).

to n_2 if the absolute value of the difference, $|\tilde{n}_2 - n_2|\hbar$, is smaller than the volume of the elementary cell in phase space (i. e., $\hbar/2$ for each degree of freedom). The transition probability $P_{n_2 n_1}$ for the process $n_1 \rightarrow n_2$ is determined as that part of the total number of trajectories leading from state n_1 to n_2. Evidently, this enables only the calculation of classically allowed transitions and does not satisfy, besides, the principle of detailed equilibrium because of the unsymmetric treatment of initial and final states (transition from a "point" n_1 into a "box" around n_2)[1].

On the other hand, full quantum treatments as briefly described in section 3.2. are extremely troublesome and expensive; even with additional simplifications they are at present restricted to very simple systems and models. The relatively few accurate quantum calculations carried through so far are to be considered essentially as quantitative studies of some specific effects (resonances, e.g.) and as tests for proving simpler, approximate methods which can find a broader field of application in molecular collision theory. Concerning quantum perturbation approaches, it should be mentioned that they have been employed hitherto only to a limited extent because the spectra of molecular states are rather dense and collisions usually lead to transitions having not a small probability. One important exception is found in transitions between low-lying vibrational states of diatomic molecules induced by collisions with chemically

---

[1] The principle of detailed equilibrium means equality of forward and backward transition probabilities as a consequence of the reversibility of the microscopic motions /9,11/.

Sometimes one calculates both the forward and backward probabilities, $P_{n_2 n_1}$ and $P_{n_1 n_2}$, considering the deviation of the two values as a measure of the accuracy of the method.

inert partners; in this case perturbation theory turns out
to be applicable for not too high relative translational
energies. Though not yet fully explored, the distorted-wave
approximation mentioned in section 3.2.4. seems to give merely a qualitative description of rearrangement processes.

In recent years, various attempts have been made to join
quantum and classical treatments in a way that relies as far
as possible on classical mechanics invoking quantum mechanics
just to that extent which is necessary to describe adequately
the main quantum effects in the system under consideration.
At present, most interesting and promising appear the following
concepts allowing to take into account quantum effects as well
as strong coupling between states:

- Calculation of classical trajectories (in a somewhat generalized sense) and construction of a "classical scattering
  matrix" from the actions along the trajectories. The dynamics of the collisions is described by the classical equations of motion for all considered degrees of freedom of the
  colliding particles.
  (Method of classical S matrix)

- Quantum treatment of a part of the degrees of freedom
  forming a "quantum subsystem" of the total system and classical treatment of the remaining degrees of freedom forming
  the "classical subsystem". The interaction between the quantum and classical subsystems is accounted for approximately.
  (Semiclassical approximation)

- Classical perturbation treatment of the nuclear motion or
  classical limit for the calculation of transition amplitudes between states with high quantum numbers; this
  approach may be considered as a special case of the above-
  mentioned ones.
  (Application of the correspondence principle).

To establish the connection with the rigorous quantum
formulation of section 3.2., we recall the concepts of time-
evolution and scattering operators. Given the wavefunction

$\Psi(t_1)$ of the system at some time moment t_1, the wavefunction at a later time t_2 can be generated from $\Psi(t_1)$ by a unitary operator $\hat{U}$, the so-called time-evolution operator,

$$\Psi(t_2) = \hat{U}(t_2, t_1)\, \Psi(t_1)\,, \qquad \text{(I)}$$

which can be formally written[1]

$$\hat{U}(t_2, t_1) = exp\left[-\tfrac{i}{\hbar}\, \hat{H}\cdot(t_2 - t_1)\right] \qquad \text{(II)}$$

if the system is described by a time-independent total Hamiltonian $\hat{H}$. Passing to the limits $t_1 \to -\infty, t_2 \to +\infty$ we obtain the scattering operator:

$$\hat{S} \equiv \lim_{\substack{t_1 \to -\infty \\ t_2 \to +\infty}} \hat{U}(t_2, t_1) \qquad \text{(III)}$$

(S matrix) /9,13/. The matrix elements of this operator, $S_{n_2 n_1}$, taken between the internal quantum states n_1 and n_2 of the system before and after the collision, respectively, give in a direct way the scattering amplitudes $f_{n_2 n_1}$ introduced in section 3.2.3., eqs. (3.2.-46,47)[2], compare also /9,13/.

---

[1] Sometimes this operator is called the dynamical operator or the propagator or the Green's function.

[2] As in section 3.2.3., the indices n_1, n_2 (denoted there by a and b, respectively) mean collective quantum numbers for the internal states in the initial and final channels.

The fundamental feature of quantum-mechanical scattering theory is to calculate transition-probability a m p l i - t u d e s , $S_{n_2 n_1}$, the square modulus of which gives the transition p r o b a b i l i t y of the process,

$$P_{n_2 n_1} = \left| S_{n_2 n_1} \right|^2 \qquad \text{(IV)}$$

(thus taking care of the quantum-mechanical superposition principle), rather than to determine the transition probability directly as classical mechanics does (see section 3.1.).

It is possible to incorporate quantum aspects into a basically classical treatment by retaining the concept of dealing with amplitudes but evaluating these amplitudes in some approximate way taking classical limits (section 4.1.). In the semiclassical approach, on the other hand, the motion in all those degrees of freedom for which quantum effects are assumed to play an essential role, is treated quantum-mechanically (section 4.2.). By applying perturbation theory and the correspondence principle, respectively, in this approximation both approaches are seen to become identical (section 4.3.).

If not stated otherwise, the discussions of this chapter are confined to electronically adiabatic processes. To avoid confusion, table 4.-1 gives an overview on terminology and on connections between the different methods.

4.1. Classical S-Matrix Method

The most straightforward concept to deal with quantum effects within the general framework of classical mechanics is to pass to the classical limit of the complete quantum S matrix of the collisional problem /30/. For simplicity, we consider adiabatic processes with given potential; the theory can be generalized, however, to non-adiabatic processes /30/.

A typical problem in molecular scattering theory is to find the transition probability between asymptotic eigenstates of a molecular system before and after collision (e.g. $A + BC$

Table 4.-1: Different Approaches to the Calculation of Transition Probabilities in Adiabatic Molecular Collisions

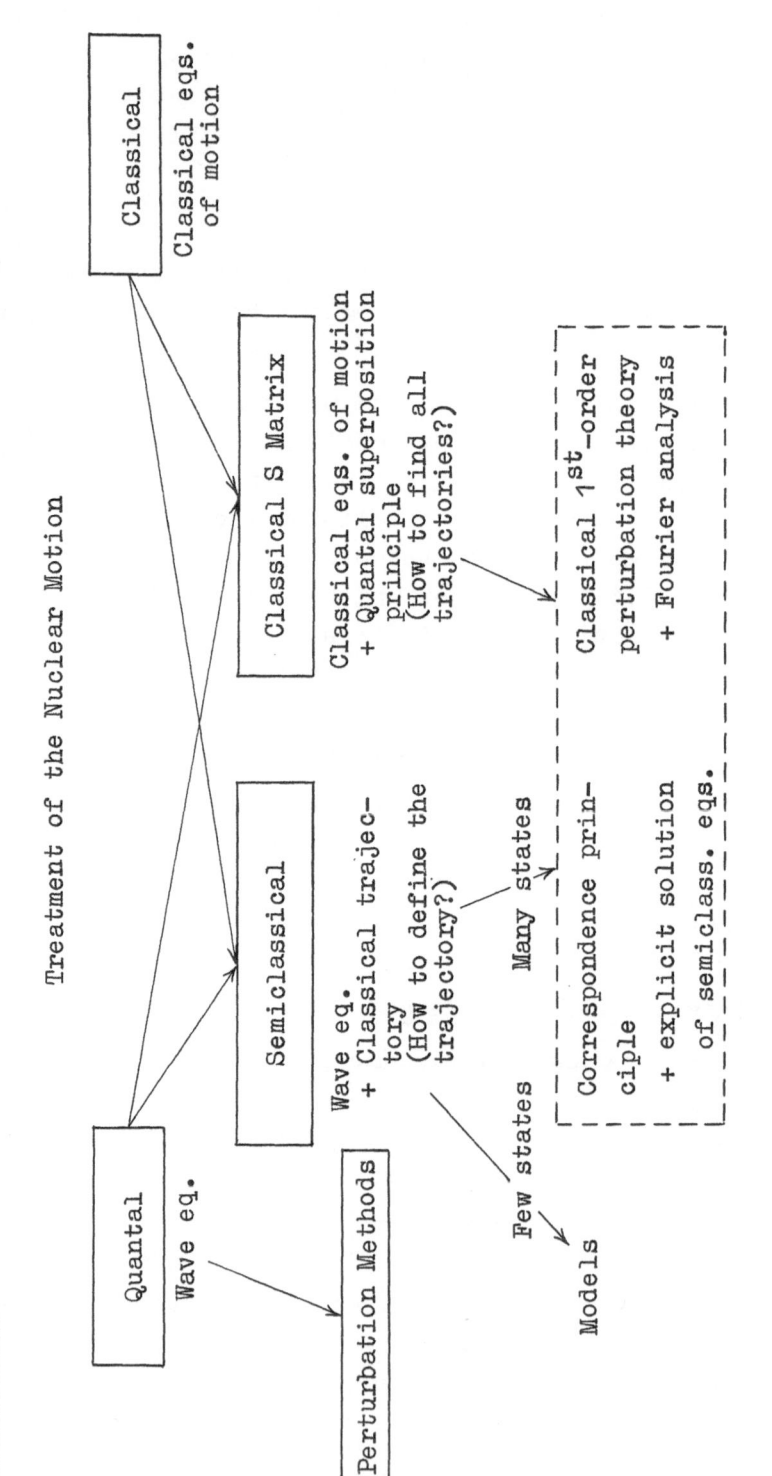

and $AB + C$, respectively) characterized by the relative momenta P, P' of the motion $A - BC$ and $AB - C$, respectively, and some quantum numbers n, n' for the internal motion of the stable, non-interacting molecules BC and AB, respectively. These asymptotic states are described by wavefunctions of the form (3.2-45),

$$|Pn\rangle = e^{iPR/\hbar} \chi_n ,\qquad (1)$$

being eigenfunctions of the Hamiltonians $\hat{H}_o$ and $\hat{H}_o'$ in the initial and final channels[1], respectively. The probability amplitude of the process $P, n \rightarrow P', n'$ is given by the matrix element

$$\widetilde{S}_{n'n} =$$

$$\lim_{\substack{t_1 \rightarrow -\infty \\ t_2 \rightarrow +\infty}} \langle P'n' | e^{\frac{i}{\hbar}\hat{H}_o' t_2} e^{-\frac{i}{\hbar}\hat{H}(t_2 - t_1)} e^{-\frac{i}{\hbar}\hat{H}_o t_1} | Pn \rangle \qquad (2)$$

($\hat{H}$ is the total Hamiltonian) of an operator which represents the limit $t_1 \rightarrow -\infty$, $t_2 \rightarrow +\infty$ of the propagator (II) with correctly separated unperturbed time dependences. This expression can be written

$$\widetilde{S}_{n'n} = \lim_{\substack{t_1 \rightarrow -\infty \\ t_2 \rightarrow +\infty}} e^{\frac{i}{\hbar}E(t_2 - t_1)} \langle P'n' | e^{-\frac{i}{\hbar}\hat{H}(t_2 - t_1)} | Pn \rangle \qquad (3)$$

$$= \delta(E' - E) S_{n'n}(E) . \qquad (3')$$

---

[1] In section 3.2. for the multi-channel case these Hamiltonians are denoted by $\hat{H}_\alpha$ and $\hat{H}_\beta$, respectively.

The classical approximation of the S matrix, i.e. of the transition amplitude $S_{n'n}$, is obtained from eq. (3') by using the classical limits for the propagator (II) as well as for the initial and final wavefunctions (1) of the system.

To this end, definite representations of these quantities must be chosen. Let us consider, for example, the coordinate representation of the propagator (II): the matrix element $\langle x_2 | \hat{U} | x_1 \rangle$ gives the probability amplitude for finding the particles of the system at the time moment t_2 in positions collectively denoted by a coordinate x_2 if at t_1 they had been located at x_1 . Taking the classical limit[1] $\hbar \rightarrow 0$, the propagator is uniquely connected with certain characteristics of the classical trajectories corresponding to these boundary conditions $x(t_1) = x_1$, $x(t_2) = x_2$ [2]. In particular, in the coordinate representation it can be expressed as a sum over different trajectories (index j) calculated with the above-mentioned boundary conditions:

$$\langle x_2 | \hat{U}(t_2,t_1) | x_1 \rangle = \sum_j A_j \exp\left[\tfrac{1}{\hbar} \varphi_j(x_2,x_1)\right] \qquad (4)$$

where φ denotes the classical action integral (see /11/),

$$\varphi(x_2,x_1) = \int_{t_1}^{t_2} L(x,\dot{x})\,dt , \qquad (5)$$

defined as usual by means of the Lagrangian $L = 2T-H$ of the system; the A_j are some factors as yet unspecified. Ex-

[1] Practically, this limit is realized for sufficiently large particle masses.

[2] Note the difference to the "ordinary" classical-trajectory approach (section 3.1.) where a pure initial-value problem is to be solved.

pression (4) can be obtained, e.g., from FEYNMAN's "path-integral formulation" of quantum mechanics /31/ if in the limit $\hbar \to 0$ only those of all possible trajectories survive along which the action functional is stationary.

To write down the classical-limit S matrix we consider for simplicity the case of two degrees of freedom, one translational and one internal; let us imagine an electronically adiabatic collision of an atom with a diatomic molecule, $A + BC$. In the ordinary coordinate representation, using coordinate R and momentum P for the translation and coordinate r and momentum p for the internal motion, the classical limit of the propagator has a form completely analogous to (4) with phases given by eq. (5) for the corresponding Langrangian, and the classical-limit (WKB) /9/ internal wavefunction becomes

$$\chi_n(r) \simeq p^{-\frac{1}{2}} \exp\left[\frac{i}{\hbar} \int^r p(n,r)\, dr\right].\qquad(6)$$

Inserting these expressions into formula (3), one obtains the classical limit of the S-matrix element. For the present purpose, it is appropriate to describe the internal motion instead of ordinary coordinate and momentum by so-called action-angle variables /11/, n and w, n being the classical counterpart of the quantum-mechanical quantum number characterizing the internal motion[1]; passing from one set of coordinates to the other is brought about by the canonical transformations of classical mechanics /11/. Thus, leaving aside the derivation, we finally arrive at the following expression for the classical S-matrix element giving the transition amplitude from state n to state n':

---

[1] I.e., in the quantum-mechanical description, this variable will have only integer values.

$$S_{n'n} = \sum_{j} \left(\frac{1}{2\pi\hbar}\frac{\partial^2\varphi}{\partial n\,\partial n'}\right)^{1/2} exp\left[-\frac{i}{\hbar}\,\varphi_j\,(P'n',Pn)\right]; \quad (7)$$

it is determined by the action φ along the possible classical trajectories (numbered by the index j) connecting
states n and n' [1], i.e. corresponding to the boundary
conditions

$$n(t_1) = n \quad , \quad P(t_1) = P,$$

$$n(t_2) = n' \quad , \quad P(t_2) = P'. \tag{7a}$$

The generalization to many degrees of freedom will not be
discussed here /30/.

As can be seen from these considerations, the application
of the "classical S-matrix" approach requires the numerical
integration of the classical equations of motion. The present
problem, however, is different from that in the ordinary
quasiclassical trajectory method where the i n i t i a l
v a l u e s of the dynamical variables are given: trajectories are to be found which lead, according to the b o u n d -
a r y c o n d i t i o n s (7a), from a given initial n to
a given final n' . Each trajectory which satisfies these
conditions corresponds to a certain initial phase[2] (i.e.
initial value of the angle variable) being frequently not
unique - several trajectories connect states n and n' .

---

[1] This expression (7) is valid if the difference of the
phases of different trajectories is sufficiently large.

[2] In multi-dimensional problems, we have a set of variables n
("quantum numbers") and, correspondingly, a set of angle
variables with initial phases.

The practical determination of such trajectories which are defined "from both ends" is a rather difficult task. Along each trajectory j of this kind the classical action integral φ_j , eq. (5), is to be evaluated giving according to expression (7) directly the phase of the corresponding contribution and, via the derivatives with respect to n and n' , the amplitude factor $A^j_{n'n}$ determining the classical probability $P^j_{n'n} = |A^j_{n'n}|^2$ for trajectory j .

The transition probability is then, according to eq. (IV), obtained in the form

$$P_{n'n} = \sum_j P^j_{n'n} + \sum_{j \neq i} \sum (P^j_{n'n})^{1/2} (P^i_{n'n})^{1/2} \cos\left\{\frac{1}{\hbar}(\varphi_j - \varphi_i)\right\}. \quad (8)$$

This expression[1] takes interference effects into account, represented by the second sum in eq. (8) which appears as a consequence of the quantum superposition principle. If these terms are neglected (which is justified, e.g., in case the second sum oscillates quickly for small variations of the trajectory parameters), eq. (8) simplifies to

$$P_{n'n} = \sum_j P^j_{n'n} \quad (9)$$

representing the classical limit of the classical S-matrix method. It should be noted that, in contrast to the ordinary quasiclassical probabilities (section 3.1.), the classical probabilities (8) and (9) satisfy the principle of detailed equilibrium /9,13/ because of the symmetric treatment of initial and final states.

In many cases, interference effects may be neglected but there exists a series of phenomena for which they are basically important: for example, they explain in classical terms

---

[1] Sometimes it is called the p r i m i t i v e classical S-matrix approach.

the fulfillment of the rigorous selection rule ΔJ = integer for rotational transitions in homonuclear diatomic molecules and the non-monotonous dependence of the vibrational transition probability on quantum numbers and energy.

Formulae (8) and (9) refer to classically allowed transitions, i.e. transitions $n \rightarrow n'$ for which exists at least one trajectory connecting these states. In a more general formulation /30/, the classical S-matrix method can be extended also to classically forbidden transitions. In the framework of the primitive version of the method as given here, this problem requires finding complex-valued trajectories and phases for the transition $n \rightarrow n'$, i.e. the integration of the classical equations of motion along a complex-valued time contour which is chosen in a way that the trajectory is "forced" to pass from n to n'. This procedure leads to the following formula for the transition probability:

$$ P_{n'n} = \frac{1}{2\pi} \left| \frac{\partial^2 \varphi}{\partial n \, \partial n'} \right| \exp \left\{ - \mathcal{Im} \, \frac{2}{\hbar} \, \varphi \right\} \qquad (10) $$

where from all possible trajectories one is retained having the minimal imaginary part of the action. In this expression, being valid only for sufficiently large $\mathcal{Im} \, (\varphi/\hbar)$, initial and final phases may be complex-valued.

To illustrate the present state of application of the classical S-matrix method let us consider the collinear symmetric reaction

$$ H + H_2 (\nu = 0) \longrightarrow H_2 (\nu = 0) + H . \qquad (11) $$

In fig. 4.-1, the probability of this process in the threshold-energy region as calculated by the classical S-matrix method /32/ (dashed line) is compared with accurate quantum results (full line) and with quasiclassical "point-to-box" transition probability (dash-dotted line). It is seen that tunneling[1] which is accounted for by the classical S-matrix calculation makes a large contribution (see also section 3.2.4., fig. 3.-21) and that the classical S-matrix method is capable of giving very accurate results.

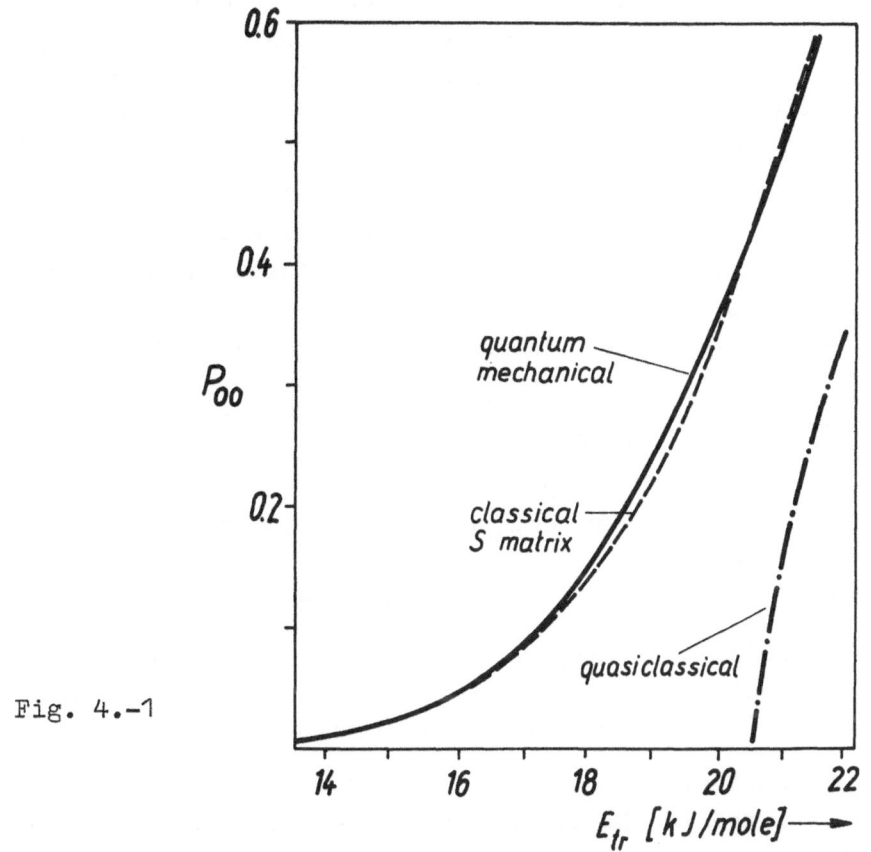

Fig. 4.-1

---

[1] In the case of motions in more than one degree of freedom (see section 3.2.3.1.), it is difficult to define tunneling in a unique way. By tunneling one may mean, for example, the onset of a process below the threshold obtained in a classical calculation (accurate trajectory treatment) or below the effective potential barrier (0.127 eV $\hat{=}$ 12,3 kJ/mole for the process (11), see also mark B in fig. 3.-21) taking into account the reactant internal energy (0.269 eV for ground-state H_2).

Formulae (8) and (10) are correct only sufficiently far from the boundary separating the regions of classically allowed and classically forbidden transitions; this boundary is defined by the condition that two or more trajectories connecting one and the same pair of states n and n', are very close together or (in the limit) merge. For building up the classical S matrix in cases where the phase difference $(\varphi_j - \varphi_i)/\hbar$ is not large, the treatment must be based on the general integral representation of the S matrix taking into account not only the contributions of a finite number of trajectories to the scattering amplitude but considering a continuous manifold of trajectories. A satisfactory result frequently can be obtained by using so-called uniform approximations /30/. In case of two closely neighboured trajectories 1 and 2, for example, the most widely known approximation of $P_{n'n}$ is given by means of the Airy functions Ai and Bi :

$$P_{n'n} = \left| \sqrt{P_{n'n}^1} + \sqrt{P_{n'n}^2} \right|^2 \pi z^{1/2} Ai(-z)$$

$$+ \sqrt{P_{n'n}^1 P_{n'n}^2} \ \pi z^{1/2} Bi(-z) \tag{12}$$

$$z \equiv \left| 3(\varphi_1 - \varphi_2)/4\hbar \right|^{2/3}.$$

In the two limiting cases $z \gg 1$ and $z \ll 1$, expression (12) becomes identical with (8) and (10), respectively, but allows to investigate also the intermediate cases.

We do not further discuss these problems here; instead the reader is referred to the literature (see, for example, /30,33,13f/). It should be pointed out only that the meaning and the accuracy of different uniform approximations can be understood easily by a semiclassical calculation of the S matrix in the correspondence-principle approximation (see section 4.3.): for example, the primitive approximation corresponds to an evaluation of the integral (4.3-6) using the stationary-phase method, and the Airy approximation corresponds to a representation of the action for two closely neighboured trajectories by a cubic function of the initial phase.

4.2. The Semiclassical Approach

One of the main difficulties in applying classical mechanics to molecular collision processes arises in connection with the treatment of the (quantized) vibrational and rotational motions; furthermore, if we want to deal also with electronic motion explicitly, an adequate consideration of quantum aspects is unavoidable. To make use of the advantages of classical mechanics as far as possible, a hybrid theory can be formulated: the so-called semiclassical or classical-path approach /31,34/.

The main idea is to separate the full set of electronic plus nuclear degrees of freedom into two groups - quantum ("internal") degrees of freedom described by coordinates q , and classical ("external") ones described by coordinates Q . The total Hamiltonian of the system can be written

$$H(P,Q;p,q) = H(P,Q) + H(p,q) + V(Q,q) \qquad (1)$$

where P and p represent the set of conjugate generalized momenta corresponding to Q and q , respectively; it is assumed that $V(Q,q) \to 0$ before and after the collision.

The classical subsystem will be described by trajectories $Q(t)$, the quantum subsystem by wavefunctions $\Psi(q,t)$. To find these quantities, we start from the integral representation of the propagator with respect to the quantum numbers n_1 and n_2 of the internal motion (i.e., the motion of the quantum subsystem):

$$U_{n_2 n_1}(Q_2 t_2, Q_1 t_1) = \langle \phi_{n_2}(q_2) | U(Q_2 q_2 t_2, Q_1 q_1 t_1) | \phi_{n_1}(q_1) \rangle ; \quad (2)$$

here $U(Q_2 q_2 t_2, Q_1 q_1 t_1)$ designates the propagator in the coordinate representation, $\phi_{n_1}(q)$ and $\phi_{n_2}(q)$ being stationary eigenfunctions of the Hamiltonian $\hat{H}(p,q)$:

$$\hat{H}(p,q) \, \phi_n(q) = E_n \, \phi_n(q). \quad (3)$$

Expression (2) can be rewritten in form of a "path integral" of the classical subsystem,

$$U_{n_2 n_1}(Q_2 t_2, Q_1 t_1) = \int d[\tilde{Q}] \exp\left\{ \frac{i}{\hbar} \int_{Q_1}^{Q_2} L(\tilde{P}, \tilde{Q}) dt \right\} T_{n_2 n_1}(\tilde{Q}) , \quad (4)$$

containing the classical Lagrangian L for the classical subsystem described by the Hamiltonian $H(P,Q)$ as well as the amplitude $T_{n_2 n_1}$ for the transition of the quantum subsystem induced by the motion of the classical one along some "trajectory" $\tilde{Q} = \tilde{Q}(t)$. This transition amplitude is to be determined by solving the non-stationary Schrödinger equation for the quantum subsystem:

$$i\hbar \, \frac{\partial \Psi(q,t)}{\partial t} = \left\{ \hat{H}(p,q) + V(q, Q(t)) \right\} \Psi(q,t). \quad (5)$$

For the trajectories $Q = Q(t)$ of the classical subsystem those "trajectories" $\tilde{Q} = \tilde{Q}(t)$ are taken for which the integrand in the expression (4) corresponds to stationary phase,

$$\delta \left\{ \int_{Q_1}^{Q_2} L(\tilde{P}, \tilde{Q}) dt + \hbar \, \mathfrak{Im}(\ln T_{n_2 n_1}(\tilde{Q})) \right\} = 0. \quad (6)$$

The transition amplitude $T_{n_2 n_1}(\widetilde{Q})$ is then given by

$$T_{n_2 n_1}(\widetilde{Q}) = \int \phi_{n_2}^*(q) \exp\left(\tfrac{i}{\hbar} E_{n_2} t\right) \Psi(q,t)\, dq \qquad (7)$$

where $\Psi(q,t)$ is a solution of eq. (5) for the initial condition

$$\Psi(q,t_1) = \phi_{n_1}(q). \qquad (8)$$

It should be pointed out that, formally, the condition (6) represents some kind of Hamilton's principle of classical mechanics from which follow classical equations of motion. These equations together with equation (5) form a complete set of equations determining the trajectories of the classical subsystem and the transition amplitudes for the quantum subsystem. The coupling between these subsystems is brought about, on the one hand, by the perturbation $V(Q,q)$ acting on the quantum subsystem because of the motion of the classical one along a certain trajectory and, on the other hand, by the non-local, time- and energy-dependent effective potential the form of which is determined by the whole history of the evolution of the quantum subsystem and which occurs in the equations of motion of the classical subsystem. In this general form the problem turns out to be very complicated; it can be considered more in detail only for the most simple cases.

In the practical realization of the semiclassical approximation simplifying assumptions concerning the coupling of classical and quantum subsystems are introduced. The most widely used method is to determine the classical trajectory from the equations of motion with an effective Hamiltonian,

$$H_{\text{eff}}(P,Q,t) = H(P,Q) + V_{\text{eff}}(Q,t), \qquad (9)$$

in which the effective interaction potential V_{eff} is defined as an average,

$$V_{eff}(Q,t) \equiv \langle \Psi(q,t) | \hat{H}(p,q) + V(Q,q) | \Psi(q,t) \rangle, \quad (10)$$

of all q-dependent parts of the total Hamiltonian (1) with the instantaneous state $\Psi(q,t)$ of the quantum subsystem[1]. If $\Psi(q,t)$ is expanded in terms of the eigenfunctions $\phi_n(q)$ of the Hamiltonian $\hat{H}(p,q)$, eq. (3),

$$\Psi(q,t) = \sum_n a_n(t) \exp(-\frac{i}{\hbar} E_n t) \phi_n(q), \quad (11)$$

then the complete set of equations for the trajectories $Q = Q(t)$ and the amplitudes $a_n(t)$ becomes

$$\dot{Q} = \frac{\partial}{\partial P} H_{eff}(P,Q,t) \quad , \quad \dot{P} = -\frac{\partial}{\partial Q} H_{eff}(P,Q,t), \quad (12)$$

$$i \dot{a}_n = \sum_m \frac{1}{\hbar} V_{nm}(t) \exp\left\{\frac{i}{\hbar}(E_n - E_m)t\right\} a_m \quad (13)$$

where

$$V_{eff}(Q,t) = \sum_n |a_n|^2 (E_n + V_{nn}) + \sum\sum_{n \neq m} a_n^* a_m V_{nm}(t) e^{-\frac{i}{\hbar}(E_m - E_n)t}, \quad (14a)$$

$$V_{nm}(t) \equiv \langle \phi_n(q) | V(q, Q(t)) | \phi_m(q) \rangle. \quad (14b)$$

It can be shown that in this effective-potential approxima-

---

[1] Notice the formal analogy to the adiabatic approximation (see section 2.2.2.).

tion the average value of the total energy of the system is conserved, i.e.

$$\frac{d}{dt} \langle \Psi | \hat{H}(P,Q;p,q) | \Psi \rangle = 0.$$

(15)

In the framework of the more simple, so-called external-force approximation, the effective potential V_{eff} is assumed to be time-independent and can therefore be incorporated into $H(P,Q)$. As the first step, the classical problem is solved and the trajectories found are then used for solution of the quantum problem (5) or (13). In this approximation the average value of the total energy is not conserved because in the equations the influence of the quantum system back onto the classical one does not appear; instead only the energy of the classical subsystem is conserved. This defect of the external-force approximation, in many cases, seems not to be essential. The main shortcoming of this approximation as well as of the effective-potential approximation is the inaccurate consideration of the **reverse** coupling being important even in the limit of small interaction $V(Q,q)$. In fact, if all amplitudes a_n except one, the initial a_{n_1}, are small then V_{eff} with high accuracy is equal to $V_{n_1 n_1}$ and both approximations practically give the same result. This, however, is not correct since the probability $P_{n_2 n_1}$ (let it be small) has been calculated for a trajectory which does not feel the (possibly large) change of the quantum subsystem.

Let us consider, as an example, a system with two degrees of freedom - the collinear collision of an atom with a diatomic molecule in the adiabatic approximation, given the interaction potential. The internal (vibrational) motion of the molecule is taken as the quantum subsystem, the translational relative motion as the classical subsystem. The latter, calculated in the external-force approximation, depends on one parameter, the energy E_Q. The transition probabilities in the quantum subsystem (transitions between vibrational states of the molecule) satisfy the usual closure relation,

$$\sum_{n_2} P_{n_2 n_1}(E_{Q_1}) = 1 \qquad (16)$$

and the reversibility condition

$$P_{n_2 n_1}(E_{Q_1}) = P_{n_1 n_2}(E_{Q_1}). \qquad (17)$$

On the other hand, the principle of detailed equilibrium (following from the time-reversibility of the motion of the total system) requires:

$$P_{n_2 n_1}(E_{Q_1}) = P_{n_1 n_2}(E_{Q_2}) \qquad (18)$$

with

$$E_{Q_1} + E_{n_1} = E_{Q_2} + E_{n_2}. \qquad (19)$$

Evidently, equations (17) and (18) are contradicting each other. One can try to remove this contradiction by noticing that the parameters of the classical trajectory used for the calculation of the transition probability $n_1 \rightarrow n_2$, on principle are determined with some uncertainty. In particular, using the exact conservation laws one can find the connection of initial and final parameters of a trajectory for a given transition $n_1 \rightarrow n_2$. Then, in eq. (17) instead of E_{Q_1} some energy E_{12} lying between E_{Q_1} and E_{Q_2} can be inserted. Such symmetriza-tion of parameters of a trajectory may be supposed to improve the result; various comparisons of semiclassical with quantum calculations in fact show that this proves right. However, such procedure leads to new difficulties connected with the violation of the closure condition (16) for symmetrized trajectories.

As can be seen from these considerations, the semiclassical approximation suffers from intrinsic contradictions when not rigorously carried through; using the results of such approach one has to ascertain that a calculated effect lies outside of the range of uncertainty of the method itself. We mention two practically important cases in which the application of the semiclassical approach should not be dubious:

1. Small transition probabilities: $P_{n_2 n_1} \ll 1$ for $n_2 \neq n_1$. In this case symmetrization does not violate the closure condition which is trivially satisfied because of $P_{n_1 n_1} \approx 1$.

2. Small changes of trajectory parameters: $E_{Q_2} \approx E_{Q_1}$. In this case one can renounce symmetrization supposing that $P_{n_2 n_1}(E_{Q_1})$ and $P_{n_1 n_2}(E_{Q_2})$ differ only slightly.

Since the basic difficulty in applying the semiclassical approximation practically is the solution of the non-stationary wave equation (5), analytically solvable models evidently play an important role.

In the first line, the so-called two-state models /35/ must be mentioned which are appropriate to describe transitions between two states of the quantum subsystem provided that transitions to all remaining states have small probabilities. Such two states may be two accidentally closely neighbouring vibrational states of a polyatomic molecule or two electronic states in a region where the electronic terms approach each other or intersect (see chapter 5.).

Another important model treats the vibrations of a quantum oscillator as induced by a time-dependent force with variable frequency. The solution of this problem has been studied quite thoroughly (see, e.g., /36/) and the transition probabilities are given by expressions containing special functions. Here we only briefly mention that the transition probabilities can be found without actual solution of the wave equation using the fact that the Green's function of such system (i.e., the propagator) can be expressed by the classical action (see section 4.1.). The solution of this problem is widely used in the theory of vibrational relaxation of molecules.

4.3. Calculation of Transition Probabilities Using the Correspondence Principle

Utilizing specific features of the interaction between states with large quantum numbers, a simplified semiclassical approximation can be formulated.

We consider transitions between states n and n' for both $n, n' \gg 1$; furthermore we suppose that all final states n' to which transitions occur are sufficiently close to the initial state n – i.e. $|n'-n| \ll n, n'$. In this limiting case the matrix element $V_{n'n}(Q) = \langle \phi_{n'} | V(q,Q) | \phi_n \rangle$ appearing in eqs. (4.2.–12,13) can be set, according to the correspondence principle, equal to the Fourier component of the classical quantity $V(q(t), Q)$. In particular, for one degree of freedom we have:

$$ V_{n+s,n}(Q) = \frac{1}{T_{\bar{n}}} \int_0^{T_{\bar{n}}} V(q_{\bar{n}}(t), Q) e^{i\omega_{\bar{n}}st} dt \quad , \atop (n \gg s) \tag{1} $$

here $q_{\bar{n}}(t)$ designates a classical trajectory of the unperturbed periodic motion corresponding to some state "intermediate" between the quantum states n and $n+s$, $\omega_{\bar{n}} = 2\pi/T_{\bar{n}}$ the frequency. In the limit $n \gg s$ the matrix element $V_{n+s,n}$ strongly depends on s but weakly on n . If one neglects the latter dependence completely, i.e. assumes $V_{n+s,n}$ to depend on s only, then the (in this case infinite) matrix $V_{n'n}$ takes the specific form of a matrix with elements depending only on the difference of the indices. With the same degree of accuracy one may suppose that in the neighbourhood of state n the energetic spectrum of the unperturbed system is equidistant. Then the set of equations (4.2.–13) for the transition amplitudes simplifies to

$$i \dot{a}_{n'} = \frac{1}{\hbar} \sum_m V_{|m-n'|}(Q(t)) e^{i\omega(m-n')t} a_m . \tag{2}$$

Since the matrices $W_{km}(t) = \frac{1}{\hbar} V_{|k-m|}(t) \exp(i\omega(k-m)t)$ on the right-hand side, taken at different time instants, commute[1]), the solutions of eqs. (2) when expressed by the S matrix have the form

$$S_{n'n} = \left\{ \exp\left(-i \int_{-\infty}^{\infty} W dt\right) \right\}_{n'n} . \tag{3}$$

An explicit formula is obtained by realizing that the eigenvectors of the matrix W have the form $x_n(w^\circ) \sim \exp(inw^\circ)$ and the corresponding eigenvalues $\mu(w^\circ)$ can be written

$$\mu(w^\circ) = \sum_{n,m} x_n^*(w^\circ) W_{nm} x_m(w^\circ) = \frac{1}{\hbar} \sum_s V_s(Q) e^{i\omega st - isw^\circ} . \tag{4}$$

Taking into account that the last sum in this equation can be condensed into the classical function

$$\sum_s V_s(Q) e^{i\omega st - isw^\circ} = V\left[q(t - \frac{w^\circ}{\omega}); Q\right] , \tag{5}$$

we finally find:

$$S_{n+s,n} = \frac{1}{2\pi} \int_0^{2\pi} dw^\circ \exp\left\{ isw^\circ - \frac{i}{\hbar} \int_{-\infty}^{\infty} V\left[q_{\bar{n}}(t - \frac{w^\circ}{\omega_{\bar{n}}}), Q(t)\right] dt \right\} . \tag{6}$$

The indices $\bar{n}$ at q and ω indicate that integration over

---

[1])Arbitrary matrices with elements depending solely on the difference of the indices, commute.

t is to be performed along some average trajectory; in the limit $\bar{n} \gg s$ the choice of this trajectory only weakly influences the result. As can be seen from expression (6), w° has the meaning of the initial phase of the motion in coordinate q . Furthermore, from the integral representation (6) the closure and symmetry relations for the probability ,
$P_{n+s,n}(\bar{n}) = |S_{n+s,n}(\bar{n})|^2$, can be obtained easily.

Formula (6) can be generalized to more (N) degrees of freedom of the quantum subsystem. Let $\underline{s}$ denote the set of changes of the quantum numbers $\underline{n}$ ($\underline{s} \equiv \{s_1, s_2, \ldots, s_N\}$) and $\underline{w}^{\circ}$ the initial phases of the motion in the corresponding classical coordinates; for convenience, let us assume also that V depends on the canonical variables $\hbar\underline{n}$ and $\underline{w}$, i.e.
$V[q, Q] = V[n_1, \omega_1 t - w_1^{\circ}; n_2, \omega_2 t - w_2^{\circ}; \ldots; Q(t)]$. Then the expression

$$S_{\underline{n}+\underline{s},\underline{n}} = \frac{1}{(2\pi)^N} \int d^N \underline{w}^{\circ} \exp\left\{i\underline{s}\,\underline{w}^{\circ} - \frac{i}{\hbar}\int V[\bar{n}, \underline{w}(\bar{n}, t, \underline{w}^{\circ}), Q(t)]dt\right\} \quad (7)$$

holds. We notice an important **property** of the S matrix following from this representation (7). If only a so-called partially-averaged transition probability is needed, i.e. the probability summed over all final states of some degrees of freedom, this quantity can be obtained without calculation of the complete probability. Let the total set of quantum numbers $\underline{n}$ split into two groups, $\underline{n}_a$ and $\underline{n}_b$, and we are interested only in the transition probability between states which are different in the quantum numbers $\underline{n}_a$,

$$P_{\underline{n}'_a \underline{n}_a} = \sum_{\underline{n}'_b} P_{\underline{n}' \underline{n}} . \quad (8)$$

Using the property $(2\pi)^{-1} \sum_l \exp[il(w^{\circ} - \tilde{w}^{\circ})] = \delta(w^{\circ} - \tilde{w}^{\circ})$ of the sum, we find

$$P_{\underline{n}'_a\,\underline{n}_a} = \frac{1}{(2\pi)^{N_b}} \int d^{N_b}\underline{w}^o_b \; P_{\underline{n}'_a\,\underline{n}_a}(\underline{w}^o_b) \qquad (9)$$

where

$$P_{\underline{n}'_a\,\underline{n}_a}(\underline{w}^o_b) = \left| S_{\underline{n}'_a\,\underline{n}_a}(\underline{w}^o_b) \right|^2, \qquad (9a)$$

$$S_{\underline{n}'_a\,\underline{n}_a}(\underline{w}^o_b) = \frac{1}{(2\pi)^{N_a}} \int d^{N_a}\underline{w}^o_a \; \exp\Big\{ i\,(\underline{n}'_a - \underline{n}_a)\,\underline{w}^o_a$$

$$-\frac{i}{\hbar}\int V\big[\bar{\underline{n}}_a, \bar{\underline{n}}_b, \underline{w}_a(t,\underline{w}^o_a), \underline{w}_b(t,\underline{w}^o_b); Q(t)\big]dt\Big\}. \quad (9b)$$

In this way, the partially-averaged transition probability is determined by averaging the transition probability $P_{\underline{n}'_a\,\underline{n}_a}(\underline{w}^o_b)$ for fixed initial phases of those degrees of freedom which we are not interested in.

The expression (6) is formally analogous to the integral representation of the classical S matrix if for evaluation of the phase φ classical perturbation theory is used. The integral in the exponent of eq. (6) is equal to the first-order correction $A^{(1)}$ to the action, due to the additional interaction V. If we choose as the zeroth-order approximation a trajectory corresponding to the unperturbed Hamiltonian and to the initial conditions given by the canonical coordinates $\hbar\underline{n}$ and $\underline{w}^o$, we obtain

$$A^{(1)}(\underline{n},\underline{w}^o) = \int V\big[\underline{n}, \underline{w}(t,\underline{w}^o); Q(t)\big]dt . \qquad (10)$$

Knowledge of the function $A^{(1)}(\underline{n},\underline{w}^o)$ allows a calculation of the change of the classical variables $\underline{n}$ in the collision:

$$n_k' - n_k = \frac{1}{\hbar} \frac{\partial}{\partial w_k^o} A^{(1)}(n_1, \ldots, n_N; w_1^o, \ldots, w_N^o) \; ; \qquad (11)$$

this is useful in many cases for determining the limits of changes of n_k in the classically accessible region and for calculation of the probability in the classical version of the classical S-matrix method (approximation of the type (4.1.-9)).

As an example of calculation of transition probabilities using the correspondence principle we consider the forced quantum harmonic oscillator. This model is widely employed in the theory of vibrational-translational energy exchange /36/ and seems also to be useful for the interpretation of the energy redistribution in exothermic exchange reactions.

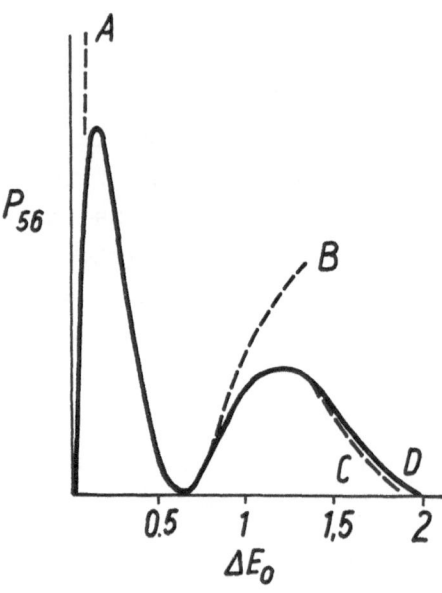

Fig. 4.-2

For the forced harmonic oscillator the **exact** transition **probability** depends on one parameter, ΔE_o , the energy transferred by the force to the classical oscillator being initially at rest. Fig. 4.-2 shows results of several approximations for the probability of the transition between vibrational states $6 \rightarrow 5$: curve A corresponds to the first-order q u a n t u m perturbation treatment, curve B represents a multi-state solution of the non-stationary Schrödinger equation (4.2.-5) with 16 states of expansion (4.2.-11) being taken into account, and curve C is calculated using the correspondence principle. It is seen that the latter which is based on the first-order c l a s s i c a l perturbation treatment provides a good approximation to the exact semiclassical result (curve D).

Concluding we point out that the condition of applicability of classical perturbation theory (weak perturbation of the trajectory by the additional interaction) underlying formula (6) and leading to the requirement of small quantum-number differences $n_k' - n_k$ compared with their values themselves, permits the description of multi-quantum transitions. This gives us the possibility to go beyond the framework of quantum-mechanical perturbation theory which is reliable only for small transition probabilities. Being a variant of the semiclassical approximation, this method allows some improvements by means of an optimal choice of the zeroth-order trajectory, in particular through a symmetrization of its initial and final parameters. Concerning this, some recent success is to be mentioned in the general formulation of the method aiming at a more complete account of the interaction of the subsystems as well as in its concrete application to problems of vibrational /37/ and rotational /38/ excitation of molecules. A comparison of the results with accurate quantum calculations shows that evaluation of probabilities according to formulae (7) and (9) can be recommended for a further calculation of cross sections and rate constants of concrete processes. Of course, if the interaction V is so strong that classical perturbation theory cannot be applied, for the determination of the S matrix classical trajectories must be used, i.e. one has to resort   to the classical S-matrix formalism in full extent.

5. Theory of Non-Adiabatic Transitions in Atomic and Molecular Collision Processes

The basic assumption that enables one to introduce the concept of potential energy for the motion of a system of atoms is the adiabatic approximation (see section 2.2.2.). According to this approach, the allowed energy levels $V_n(R)$ of the electrons are first determined for each fixed configuration R of the nuclear coordinates. If these adiabatic electronic levels are well separated from each other, that is if the Massey parameter $\gamma(R)$, eq. (2.2.-16), is sufficiently large for each relevant pair of levels, each such level can be considered as the potential energy of the nuclei or atoms belonging to the system in a given electronic state. In the adiabatic approximation one assumes that the motion of the atoms does not cause transitions between different electronic states. The problem of collisional energy transfer and rearrangement is then described in terms of the motion (classical or quantum mechanical) of atoms on a particular potential-energy surface. Transitions between different electronic states can be accounted for only by going beyond the adiabatic approximation; the main goal of the theory of non-adiabatic processes is to calculate the probabilities of such transitions.

In looking for an interpretation of the mechanism for a particular elementary process, the choice of the best approximation depends on the nature of the process and on the conditions under which it occurs. For sufficiently small kinetic energies of the atoms in the system, the adiabatic approximation often provides a satisfactory picture of the physics, and non-adiabatic effects represent a small correction only. Nevertheless, there are quite a lot of processes in which non-adiabatic transitions do play a fundamental role, and which cannot be interpreted in terms of motion along a single potential-energy surface. However, one finds frequently that tran-

sitions between electronic levels are localized within rather
small regions; in the remainder of the nuclear configuration
space adiabatic electronic terms can therefore still be thought
of as potential-energy surfaces. If this is the case, non-
adiabatic transitions may be considered in terms of atomic motion on portions of several different potential energy surfaces,
and sudden hops from one such surface to another one in regions
where the non-adiabatic interactions are strong. Such a picture
consisting of adiabatic motion interrupted by hopping becomes
less accurate with increasing kinetic energy. For sufficiently
high velocities of the atoms, regions where the adiabatic approximation is valid shrink or disappear altogether; adiabatic
electronic levels then completely lose their meaning of effective potential energy. Such a situation obtains at rather
high energies (above 100 eV, depending on the process under
consideration) so that the picture involving sudden hops is
actually useful for a wide class of elementary non-adiabatic
processes.

A fully consistent treatment of general validity, of course,
is to be based, instead of the semiclassical picture used above,
on a complete quantum-mechanical formulation as given in section 2.2.2.; after determination of the electronic energies
and wavefunctions, $V_n(R)$ and $\phi_n(\xi, R)$, the coupling operators $\hat{C}_{nn'}$ must be calculated and the set of coupled nuclear
wave equations (2.2.-9) is to be solved. Recalling the discussions in section 3.2. on the quantum-mechanical treatment
of adiabatic processes, it is obvious that the non-adiabatic
case requires a really tremendous effort. Nevertheless, first
attempts have already been made (see, for example, /39/) and
for the future we can expect also in this field at least important information on model systems.

5.1. Crossing and Pseudo-Crossing of Potential-Energy Surfaces

The criterion for the validity of the adiabatic approximation in describing the motion of atoms in the vicinity of the nuclear configuration R is that the Massey parameter, $\gamma = \Delta V \, l/\hbar u$, be large. In order to determine the regions where hopping can take place, it is therefore important to investigate where two potential-energy surfaces do in fact intersect or come close together thus leading to vanishing or small ΔV .

Qualitative considerations concerning the possible arrangements of two such surfaces are based on the symmetry of the electronic Hamiltonian $\hat{H}_{el}(R)$. This symmetry is defined by the set of transformations which when applied to $\hat{H}_{el}$, leave it unchanged; to these belong spatial transformations of the nuclear configuration (rotation, reflection etc.) as given by symmetry point groups /5/ and permutations of the spatial coordinates of electrons. The latter do not affect the non-relativistic Hamiltonian $\hat{H}_{el}$ (i.e., if $\hat{H}_{el}$ is independent on electron spin); the symmetry properties of the adiabatic eigenfunctions ϕ_n under these permutations are known to be characterized by the multiplicity of the electronic state /5/.

An appropriate analysis /9c,40/ shows that for a system of atoms with s nuclear degrees of freedom there will be the following possibilities for the behaviour of potential surfaces:

(1) If two s-dimensional potential-energy surfaces correspond to electronic states of different symmetries, they can intersect along a (s-1)-dimensional line. For a system with only one internal degree of freedom (for example, two atoms) this means that terms with different symmetries can intersect at a point, whereas for a system

with two degrees of freedom (for example, a system of three atoms with one internuclear distance fixed) the potential-energy surfaces can intersect along a certain curve.

(2) If two s-dimensional potential-energy surfaces correspond to electronic levels of the same symmetry, they can intersect along a (s-2)-dimensional line, as long as spin-orbit coupling is ignored. For a system with one degree of freedom this means that levels of the same symmetry cannot intersect, while for a system with two degrees of freedom the appropriate surfaces can touch at a point only.

(3) If two s-dimensional potential-energy surfaces correspond to electronic levels of the same symmetry in the presence of spin-orbit coupling, they can intersect only along a (s-3)-dimensional curve. For systems with one or two degrees of freedom no intersection is then possible.

On principle, the adiabatic electronic eigenfunctions $\phi_n(\xi, R)$ and the corresponding adiabatic potential-energy surfaces $V_n(R)$ can be found from the solution of the eigenvalue problem (2.2.-6):

$$\hat{H}_{el}(\xi, R)\, \phi_n(\xi, R) = V_n(R)\, \phi_n(\xi, R) . \tag{1}$$

In practice, however, one is hardly able to solve this problem exactly, and instead of ϕ_n only some approximate set ϕ_n^o is known. We can consider these ϕ_n^o and the corresponding expectation values

$$V_n^o(R) = \langle\, \phi_n^o |\, \hat{H}_{el}\, |\, \phi_n^o\, \rangle \tag{2}$$

as exact eigenfunctions and eigenvalues, respectively, of some Hamiltonian $\hat{H}_{el}^{o}$ which is different from $\hat{H}_{el}$. The set ϕ_n^o we call a "crude adiabatic basis" and treat the difference

$$\hat{\mathcal{U}} \equiv \hat{H}_{el} - \hat{H}_{el}^{o} \tag{3}$$

as a perturbation which should be taken into account at a proper stage of the calculations.

Usually $\hat{H}_{el}^{o}$ is of higher symmetry than $\hat{H}_{el}$, and the main difference between $V_n(R)$ and $V_n^o(R)$ is that produced by the reduction of symmetry. The most important effect of such a symmetry reduction in going from $\hat{H}_{el}^{o}$ to $\hat{H}_{el}$ might be, according to the above-mentioned rules, a non-crossing behaviour of two surfaces V_n and $V_{n'}$ which correspond to two crossing surfaces V_m^o and $V_{m'}^o$: crossing of V_m^o and $V_{m'}^o$ may disappear when the perturbation $\hat{\mathcal{U}}$ is taken into account. With small $\hat{\mathcal{U}}$, the non-crossing rule means that V_n and $V_{n'}$ would come close together in the region where V_m^o and $V_{m'}^o$ crossed. This situation is known as avoided crossing or pseudo-crossing of the adiabatic potential surfaces V_n and $V_{n'}$ /13/. The crucial point in deciding whether V_n and $V_{n'}$ will pseudo-cross or still cross, is to assign ϕ_m^o and $\phi_{m'}^o$ belonging to a higher symmetry group, to a particular symmetry species of the lower symmetry group.

An example of the symmetry reduction for a system of three identical atoms is given in table 5.-1 which shows in standard notations (see, e.g., /5,9c/) the symmetry properties of wavefunctions for the equilateral triangle (D_{3h}), isosceles triangle (C_{2v}), and arbitrary planar (C_s) configurations.

Table 5.-1: Correlation of symmetry operations and
            irreducible representations of point groups
            of three identical atoms

C_s			E	σ_h				
	C_{2v}		E	σ_v	C_2	σ_v'		
		D_{3h}	E	σ_h	$3C_2'$	$3\sigma_v'$	$2C_3$	$2S_3$
A'	A_1	A_1'	1	1	1	1	1	1
A'	B_1	A_2'	1	1	−1	−1	1	1
A"	A_2	A_1''	1	−1	1	−1	1	−1
A"	B_2	A_2''	1	−1	−1	1	1	−1
A'	A_1		1	1	1	1		
		E'	2	2	0	0	−1	−1
A'	B_1		1	1	−1	−1		
A"	A_2		1	−1	1	1		
		E"	2	−2	0	0	−1	1
A"	B_2		1	−1	−1	−1 ·		

The $C_{2v} \longrightarrow C_s$ reduction is illustrated in fig. 5.-1 which
presents potential surfaces for a hypothetical system of an
atom in S and P states interacting with a closed—shell homonuclear molecule of fixed internuclear distance; the coordinates are the distance R from the atom to the center of the
molecule and the angle θ characterizing the distortion of the
isosceles triangle. The two A' terms cross at a point (conic
or "funnel" crossing); the A' and A" terms cross along a line.

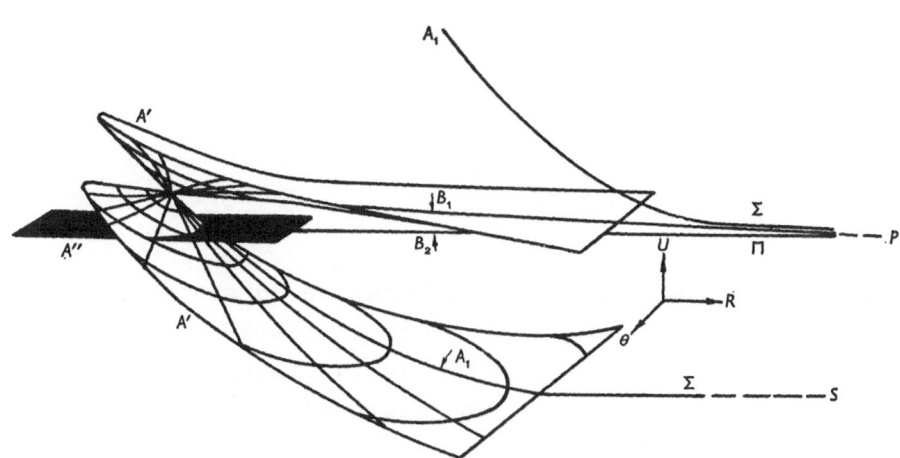

Fig. 5.-1

5.2. Non-Adiabatic Coupling and Selection Rules

In the semiclassical approximation (section 2.2.2.), the electronic wavefunction $\Phi(\xi,t)$ is to be determined from the time-dependent Schrödinger equation

$$\hat{H}_{el}\, \Phi(\xi,t) = i\hbar\, \frac{\partial}{\partial t}\, \Phi(\xi,t). \tag{1}$$

If we expand Φ , analogously to eq. (2.2.-14), in terms of the crude adiabatic basis ϕ_n° ,

$$\Phi(\xi,t) = \sum_n a_n^\circ(t)\, \phi_n^\circ(\xi, R[t])\, exp\left\{-\frac{i}{\hbar}\int^t V_n^\circ\, dt\right\} , \tag{2}$$

we obtain the following set of equations for the unknown coefficients:

$$i\hbar\, \dot{a}_n^\circ = \sum_m \left(C_{nm} + U_{nm}\right) exp\left\{-\frac{i}{\hbar}\int^t (V_m^\circ - V_n^\circ)\, dt\right\} a_m^\circ \tag{3}$$

with

$$C_{nm} = \langle \phi_n^o \, | -i\hbar \frac{\partial}{\partial t} | \, \phi_m^o \rangle , \qquad\qquad (3a)$$

$$\mathcal{U}_{nm} = \langle \phi_n^o \, | \, \hat{\mathcal{U}} \, | \, \phi_m^o \rangle . \qquad\qquad (3b)$$

From the right-hand side of this equation we see that the coupling between states of the crude adiabatic basis is due to the dynamic interaction $\hat{C} = -i\hbar\partial/\partial t$ induced by the motion of the nuclei, and by the static interaction $\hat{\mathcal{U}}$ which is not taken into account in H_{el}^o .

Two limiting cases can be distinguished. If the crude adiabatic functions ϕ_n^o provide a good approximation to the exact adiabatic eigenfunctions ϕ_n , the interaction matrix elements $\mathcal{U}_{nm}$ are small and we can neglect them; expansion (2) then coincides with eq. (2.2.-14) and eq. (3) with eq. (2.2.-15). If, on the other hand, the dynamical coupling terms C_{nm} are small in comparison with $\mathcal{U}_{nm}$, then Φ is said to be represented in terms of d i a b a t i c basis functions ϕ_n^d which may, in this case, be considered independent of R .

On principle, the electronic wavefunction Φ can be expanded in terms of either set, adiabatic or diabatic, and the corresponding amplitudes, $a_n(t)$ or $a_n^d(t)$, respectively, must be determined from an infinite set of coupled equations of type (3). Truncation of the set leading to the so-called N-state approximation will give, in general, different wavefunctions Φ for the adiabatic basis and for the diabatic basis. However, if only few electronic states need to be taken into account, a proper selection of the adiabatic or diabatic basis functions often makes it possible to achieve practically identical results for either basis; this means that N func - tions are considered to give a good approximation to the complete electronic set of the problem.

The transition probabilities between the electronic states
in the crude adiabatic approximation depend, as can be seen
from eq. (3), on the matrix elements of the dynamic and static
interactions, C_{nm} and U_{nm}, respectively. Therefore, selection rules for C_{nm} and U_{nm} play an important role in
the theory of non-adiabatic transitions. Let us briefly discuss this problem separately for the dynamic and static couplings.

Dynamic Coupling C_{nm}

Since adiabatic wavefunctions are classified according to
their transformation properties under different symmetry operations of the (electronic) Hamiltonian for fixed configurations of the nuclei, it is convenient to write the transition
operator $\hat{C}$ in a form which reflects nuclear motions preserving particular symmetries. In order to clarify this question, we consider first the simplest case of two atoms. For
fixed positions of the nuclei the electronic Hamiltonian exhibits axial symmetry, so that the adiabatic electronic wavefunctions can be characterized by a quantum number Ω which
determines the component of the total electronic angular momentum along the axis of the quasimolecule /9/. The adiabatic
electronic levels of such a system depend on a single parameter, the interatomic separation R. Two adiabatic levels
have the same symmetry if they belong to the same value of
Ω ; otherwise they have different symmetry. In the first
case mentioned only pseudo-crossing is possible, while in
the second one crossing can occur.

When considering the motion of the nuclei, it is important
to note that an arbitrary displacement can be represented as a
linear combination of a radial motion that leaves the molecular axis invariant, and of a rotation which leaves the internuclear separation invariant. The operator $-i\hbar\partial/\partial t$ can
accordingly be written as a sum of operators, each describing
one of these kinds of motion (see, e.g. /13e,f/):

$$-i\hbar\, \partial/\partial t \; = -\,i\hbar\, \dot{R}\, \partial/\partial R \,-\, \omega \hat{J}_\omega \;\; ; \qquad\qquad (4)$$

here $\dot{R}$ and ω are the radial and angular velocities describing the relative motion of the nuclei, and $\hat{J}_\omega$ is the operator corresponding to the component of the orbital angular
momentum of the electrons along the angular-velocity vector
of the quasimolecule. Equation(4)enables me to express matrix
elements of the operator $-i\hbar\, \partial/\partial t$ in terms of matrix elements
of the more convenient operators $\partial/\partial R$ and $\hat{J}_\omega$ which no
longer depend on the relative velocity of the nuclei and obey
well-known selection rules. Thus matrix elements of the operator $\partial/\partial R$ vanish unless both initial and final states
have the same symmetry, while matrix elements of $\hat{J}_\omega$ vanish
unless the quantum numbers Ω of the initial and final
states differ by unity. The radial motion can therefore give
rise to non-adiabatic transitions only between electronic
levels of the same symmetry, while rotational motion can cause
transitions only between levels of different symmetries such
that the corresponding components of orbital angular momentum
along the internuclear axis differ by ± 1.

In addition to axial symmetry, a system of two atoms can
have also other symmetries (inversion, reflection in a plane
containing the internuclear axis) which lead to extra selection rules for the non-adiabatic transitions.

A summary of the selection rules in the case of two atoms
is given for dynamic coupling in the left part of table 5.-2.
Here Ω stands for the projection of the total electronic
angular momentum on the molecular axis, $\Omega = \Lambda + S_z$ with Λ
and S_z being projections of the electronic orbital angular
momentum and spin, respectively. The signs + and - at Σ terms
refer to the reflection character of the spatial component
of wavefunctions with respect to a plane containing the internuclear axis; those of the $\Omega = 0$ terms refer to the re-

flection of the total adiabatic wavefunction with spin-orbit
interaction being taken into account. Symmetry character with
respect to inversion is indicated by g and u .

Table 5.-2: Selection rules for dynamic and static
couplings in the case of two atoms

Dynamic coupling[1]	Static coupling	
	spin-orbit interaction	electrostatic (Coulomb and exchange) interaction
$\Delta\Omega = 0$ $(\mathcal{Z})$ $\Delta\Omega = \pm 1$ $(\mathcal{R})$	$\Delta S = 0, \pm 1$ $\left.\begin{array}{l}\Delta S_z = \pm 1 \\ \Delta\Lambda = \mp 1\end{array}\right\}\begin{array}{l}\Delta\Omega \\ = 0\end{array}$	$\Delta S = 0$ $\Delta\Lambda = 0$
$g \nleftrightarrow u$ $0^+ \nleftrightarrow 0^-$	$g \nleftrightarrow u$ $\Sigma^+ \longleftrightarrow \Sigma^-$	$g \nleftrightarrow u$ $\Sigma^+ \nleftrightarrow \Sigma^-$

For an arbitrary three-atomic system, there is only one
evident adiabatic quantum number, namely the reflection character of the electronic wavefunction with respect to the
plane defined by the nuclei. Rotation of this plane will in-
duce a change in reflection character, thus mixing all states
of the system.

For a particular case, the collision of an atom with a
homonuclear diatomic molecule, a more expedient classification of states may be introduced if the main contribution to
non-adiabatic coupling comes from that regions of the nuclear

---

[1]The labels $\mathcal{Z}$ and $\mathcal{R}$ denote radial and rotational motions,
respectively.

configuration space which correspond to C_{2v} symmetry. The
adiabatic quantum numbers will then be just the symbols for
the irreducible representations of this point group. In
table 5.-3 the selection rules for dynamic coupling in the
case of zero spin—orbit interaction are listed. For the C_{2v}
symmetry Z means any motion of the nuclei in the fixed plane
preserving the C_{2v} symmetry, Y is some other motion in this
fixed plane, R_z denotes a rotation about the symmetry axis,
R_x a rotation within the fixed plane, and R_y a rotation
affecting the orientation of the plane. For the C_s symmetry
Z and Y mean any motion in the plane (they can be con-
ceived as radial and tangential relative motion), R_x is
the rotation in the fixed plane, and R_z , R_y are two modes
of rotation of the plane.

Table 5.-3: Selection rules for dynamic coupling in the
case of three atoms

C_s		A'		A''	
	C_{2v}	A_1	B_1	A_2	B_2
A'	A_1	Z	Z, Y, R_x	R_z	R_y
	B_1	Z, Y, R_x	Z	R_x	R_z
A''	A_2	R_z	R_x	Z	Z, Y, R_x
	B_2	R_y	R_z	Z, Y, R_x	Z

Static Coupling U_{nm}

If we assume that the symmetry of $\hat{U}$ is the same as that of $\hat{H}_{el}$ then U_{nm} is zero unless ϕ_n^o and ϕ_m^o belong to the same symmetry species of the symmetry group of $\hat{H}_{el}$. Besides, ϕ_n^o and ϕ_m^o may be of different symmetry with respect to a higher symmetry group corresponding to $\hat{H}_{el}^o$.

For a system of two atoms the selection rules are given in the right part of table 5.-2 which shows the overall conservation of quantum numbers for the case that $\hat{H}_{el}$ and $\hat{H}_{el}^o$ are of the same symmetry ($\hat{U}$ is an axial electrostatic interaction) and the change of some quantum numbers for the case that the symmetry of $\hat{H}_{el}^o$ is higher than that of $\hat{H}_{el}$ ($\hat{U}$ is the spin-orbit coupling).

For a system of three atoms an electrostatic interaction will mix states of the same reflection character, and the spin-orbit interaction will mix all states. Thus, in the case of the latter interaction present, point and line crossings in fig. 5.-1 will become pseudo-crossings.

5.3. The Two-State Problem in Adiabatic and Diabatic Representations

The Massey criterion $\gamma \lessapprox 1$ (see section 2.2.2.) defines the region where non-adiabatic coupling might be effective. Whether it actually will be depends on the magnitude of the dynamic-coupling matrix elements C_{nm}.

However, it is important that as far as the nuclei move sufficiently slowly, the characteristic extent of such a non-adiabatic region is small. This enables one to approximate the adiabatic potential-energy surfaces and the coupling matrix elements by simple analytic functions which allow an exact solution of the corresponding equations of type (5.2.-3). Moreover, if the extent of the coupling region is small, in most cases one can restrict the consideration to

crossing or pseudo-crossing of two electronic terms only
which makes the whole problem comparatively easy.

Under these suppositions, we briefly discuss in this
section some simple models for which the probabilities of
non-adiabatic transitions can be found in a closed form, and
establish the correspondence between these models and some
representative types of elementary collision processes.

We begin with the basic set of equations for the two-state
problem formulated in the adiabatic representation (see section 2.2.2., eq. (2.2.-15)),

$$i\hbar\dot{a}_1 = C_{12} \exp\left\{\frac{i}{\hbar}\int^t (V_1 - V_2)\,dt\right\} a_2 \ ,$$

$$i\hbar\dot{a}_2 = C_{21} \exp\left\{\frac{i}{\hbar}\int^t (V_2 - V_1)\,dt\right\} a_1 \ , \tag{1}$$

assuming that the dynamic-coupling matrix element $C_{12} = \langle \phi_1 | -i\hbar\partial/\partial t | \phi_2 \rangle$ attains its maximum value at $t = 0$ and
falls off steeply on both sides of the time scale. The proba-
bility P_{12} of a non-adiabatic transition from the adiabatic
state 1 to state 2 resulting from a single passing of the
coupling region (sometimes called the one-way transition
probability), is calculated as

$$P_{12} = |a_2(t = +\infty)|^2 \tag{2}$$

by solving eqs. (1) with the initial conditions

$$a_1(-\infty) = 1 \ , \quad a_2(-\infty) = 0 \ . \tag{3}$$

The limits $t \to \pm\infty$ are understood in the sense that initially and finally the system is far from the coupling region,
i.e. $C_{12}(\pm\infty)$ is vanishingly small.

Very often it is more convenient to formulate the problem
not in the adiabatic representation but in the diabatic one
as defined in the preceding section. The change from one representation to the other is most easily achieved if we introduce a function

$$\chi(t) \equiv 2 \int_{-\infty}^{t} \frac{C_{12}}{i\hbar} \, dt \qquad (4)$$

taking on, for $t \to \mp \infty$, the limiting values

$$\chi(-\infty) = 0 \; ,$$

$$\chi(+\infty) = \text{const} \equiv \theta. \qquad (5)$$

Let us now carry out the transformation

$$\phi_1(\xi,t) = \phi_1^d(\xi) \cos \tfrac{1}{2} \chi(t) + \phi_2^d(\xi) \sin \tfrac{1}{2} \chi(t),$$

$$\phi_2(\xi,t) = - \phi_1^d(\xi) \sin \tfrac{1}{2} \chi(t) + \phi_2^d(\xi) \cos \tfrac{1}{2} \chi(t), \qquad (6)$$

consistent with relation (4) between the matrix element of
dynamic coupling and the "rotation angle" $\chi(t)$, thus de-
fining two diabatic wavefunctions $\phi_1^d(\xi)$ and $\phi_2^d(\xi)$ which
do not longer depend on time, i.e. on the nuclear configuration $R[t]$.

Instead of the expansion

$$\Phi(\xi,t) = a_1 \phi_1(\xi,t) e^{-\frac{i}{\hbar}\int^t V_1 dt} + a_2 \phi_2(\xi,t) e^{-\frac{i}{\hbar}\int^t V_2 dt} \qquad (7)$$

we may use the alternative one,

$$\Phi(\xi,t) = b_1 \, \phi_1^d(\xi) \, e^{-\frac{i}{\hbar}\int^t H_{11} \, dt} + b_2 \, \phi_2^d(\xi) \, e^{-\frac{i}{\hbar}\int^t H_{22} \, dt} \; , \qquad (8)$$

with H_{11} and H_{22} being the corresponding diagonal matrix elements of the electronic Hamiltonian $\hat{H}_{el}$ in the diabatic basis; these, and also the static-coupling matrix element H_{12} have to be found from the definition of ϕ_1^d and ϕ_2^d in terms of ϕ_1 , ϕ_2 and from C_{12} . A little algebra gives the following set of equations for the coefficients $b_1(t)$ and $b_2(t)$:

$$i\hbar \, \dot{b}_1 = H_{12} \, \exp\left\{ \tfrac{i}{\hbar} \int^t \Delta H \, dt \right\} b_2 \; ,$$

$$i\hbar \, \dot{b}_2 = H_{21} \, \exp\left\{ -\tfrac{i}{\hbar} \int^t \Delta H \, dt \right\} b_1 \; , \qquad (9)$$

with

$$\Delta H \equiv H_{11} - H_{22} = \Delta V \cos\chi \; , \qquad (9a)$$
$$(\Delta V \equiv V_1 - V_2)$$
$$2 H_{12} = \Delta V \sin\chi \; . \qquad (9b)$$

Note that the condition of localized dynamic coupling does not require that H_{12} goes to zero at both limits $t \to \pm\infty$; the ratio $2 H_{12}/\Delta H$ vanishes at $t \to -\infty$ by virtue of eq. (4) whereas at the other limit, $t \to +\infty$, this ratio tends to a constant value $tg\,\theta$.

The connection between the two representations allows us to consider two-state models either in the adiabatic basis or in the diabatic one. We shall resort mainly to the latter since it often corresponds to a zeroth-order solution of the

problem. Moreover, we will refer directly to the time dependence of ΔH and H_{12} thus leaving open the question of the
trajectory parametrization; this gives more freedom for applications to particular collision processes.

(1) The Landau–Zener Model

If we specify the quantity ΔH by a linear function and
H_{12} by a constant value,

$$\Delta H = at/t^*,$$

$$H_{12} = a, \tag{10}$$

we obtain the so-called Landau–Zener model /41/ in which the
one-way transition probability P_{12} depends on a single parameter, at^*, only.

Fig. 5.–2 shows the most typical arrangement of electronic
terms for which the Landau–Zener model commonly is employed:
$V_1(R)$ and $V_2(R)$ are two adiabatic terms, $H_{11}(R)$ and $H_{22}(R)$
the corresponding diabatic terms, C_{12} is the dynamic–coupling matrix element in the adiabatic representation, H_{12} the
static–coupling matrix element in the diabatic representation.
ΔR denotes the extent of the coupling region in the adiabatic representation, and E stands for the total energy
which defines the positions of the turning points (marked by
heavy dots) of the nuclear motion.

This model, suitable for calculating non-adiabatic transition probabilities in situations where the coupling acts within a small region near crossing or pseudo-crossing of terms,
is most widely used and will be considered in detail in the
next section. Its main limitation is that it cannot describe
transitions if the nuclear velocity is low, i.e. if the coupling occurs close to a turning point of the nuclear motion.

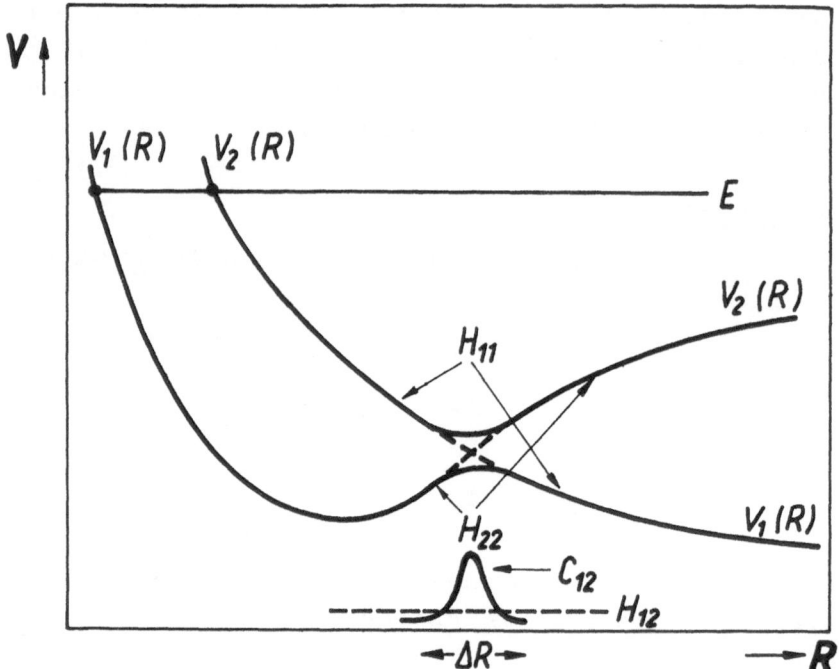

Fig. 5.-2

(2) The Extended Landau-Zener Model

To calculate transition probabilities for small nuclear velocities and even for classically "not realizable" events (i.e. if the classical trajectory does not reach the region where the dynamic coupling attains its maximum value), a simple refinement of the Landau-Zener model can be formulated /35a/. This extended Landau-Zener model is obtained by specifying the quantities ΔH and H_{12} in the following way:

$$\Delta H = a\left(\frac{t}{t^*} - \frac{b}{a}\right)^2 ,$$

$$H_{12} = a .$$

(11)

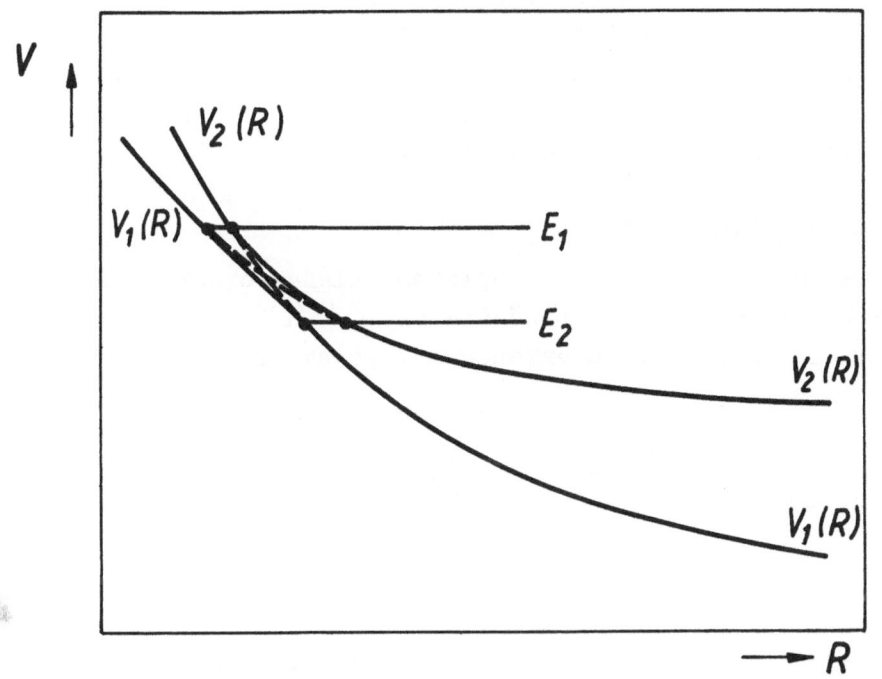

Fig. 5.-3

Fig. 5.-3 shows qualitatively a corresponding pattern of electronic terms for turning points being close to the coupling region; the total energies E_1 and E_2 define cases where the pseudo-crossing region is classically accessible or inaccessible, respectively. In the latter situation, this "classically forbidden non-adiabatic transition" is said to occur by tunneling.

The two-way transition probability (atoms approach and go away)[1] P_{12} depends on two parameters, qt^* and bt^*. With $bt^* \gg 1$, P_{12} can be expressed via the one-way Landau-Zener transition probability, eq. (5.4-12), and an interference term. If the latter is neglected (condition of rapid oscillations as discussed in section 4.1.) we get the simple result (5.4.-16) found in the next section.

(3) The Exponential Model

To take into account in a simple way higher-order corrections to the linear time dependence of $\Delta H(t)$ as adopted in the Landau-Zener model, an exponential parametrization,

$$\Delta H = a \exp(t/t^*),$$

$$H_{12} = b + c \exp(t/t^*), \tag{12}$$

has been suggested (see /35b/) which reflects, though indirectly, the exponential exchange interaction between colliding particles. By an appropriate choice of the parameters we can get from the expressions (12) either crossing or divergence of the diabatic terms. A particular term pattern (for $c = 0$) suitable for describing non-resonant charge transfer is shown

---

[1] This notion will be explained in more detail in the next section.

in fig. 5.-4. Note that the coupling region, defined by C_{12} being substantially different from zero, corresponds to the distance where the adiabatic terms begin to diverge; this is in marked contrast to the usual Landau-Zener case.

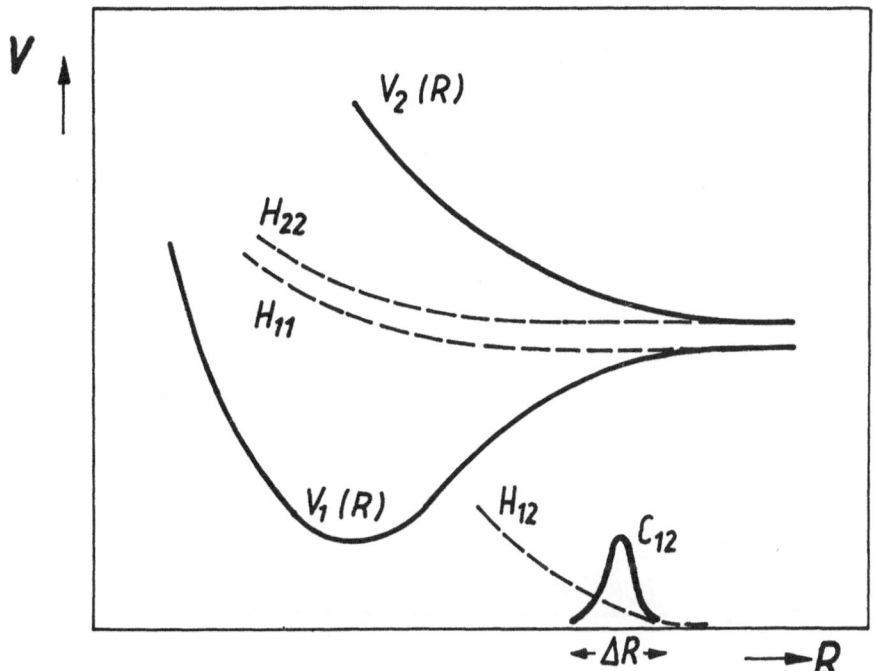

Fig. 5.-4

The one-way transition probability depends on two parameters, bt^* and a/c.

It should be noted that more general exponential models have been proposed containing more parameters and being thus more flexible /35b/; until now, however, there are no applications of these models to elementary processes.

(4) The Linear-Exponential Model

For treating cases where the diabatic terms are of long-range nature (say, Coulombic) but the interaction falls off steeply (say, exponentially), a mixed linear-exponential model may be used /35b/ with the following parametrization:

$$\Delta H = a\, t/t^{*},$$

$$H_{12} = a\, exp(-t/t_{o}), \tag{13}$$

as illustrated in fig. 5.-5 by a covalent-ionic crossing.

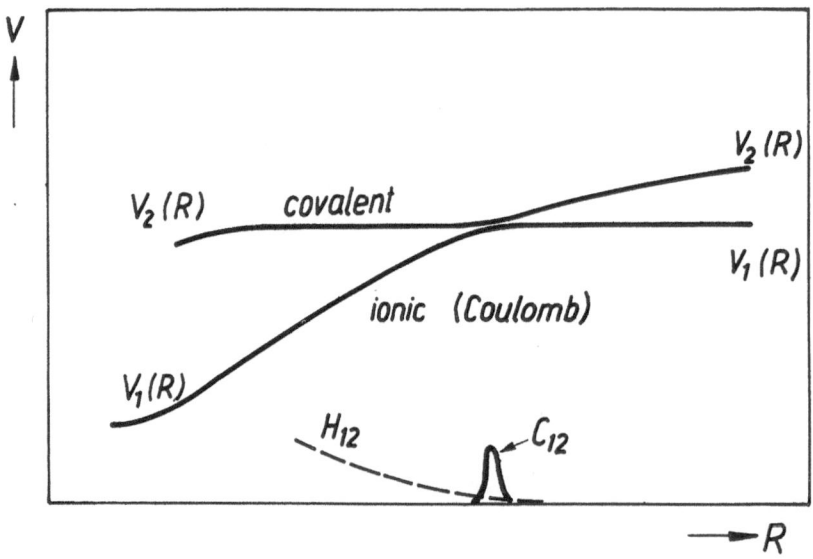

Fig. 5.-5

The one-way transition probability depends on two parameters, αt^* and αt_o ; for large t_o it reduces to the
usual Landau-Zener expression.

5.4. The Landau-Zener Model

In this section we shall examine the Landau-Zener model
specifying it to non-adiabatic coupling between electronic
terms of different and of the same symmetry. General aspects
are most transparent in the case of collisions of two atoms
which is treated here in some detail. In addition, however,
in the last part a brief description is given of the so-
called surface-hopping method as used at present in studies
of non-adiabatic chemically reactive processes.

5.4.1. Non-Adiabatic Transitions between Electronic Terms of Different Symmetry

According to the general rules discussed in section 5.1.,
two adiabatic potential-energy surfaces corresponding to different symmetry may cross. Dynamic coupling between these
electronic terms is caused by the interaction which breaks
the symmetry of the adiabatic electronic Hamiltonian, e.g.
by the rotation of the nuclear framework. This interaction
is not expected to depend critically on the nuclear configuration and it seems reasonable to assume that for two crossing surfaces the matrix element C_{12} near the crossing line
is well approximated by its value on the line, $\bar{C}_{12}$. For the
case of two colliding atoms and states 1 and 2 of different
symmetry $\bar{C}_{12}$, according to eq. (5.2.-4), is given by

$$\bar{C}_{12} = \omega \langle 1 | \hat{J}_{\omega} | 2 \rangle |_{R = R_c} \tag{1}$$

where R_c is the distance at which the crossing of the two potential curves of the diatomic system occurs.

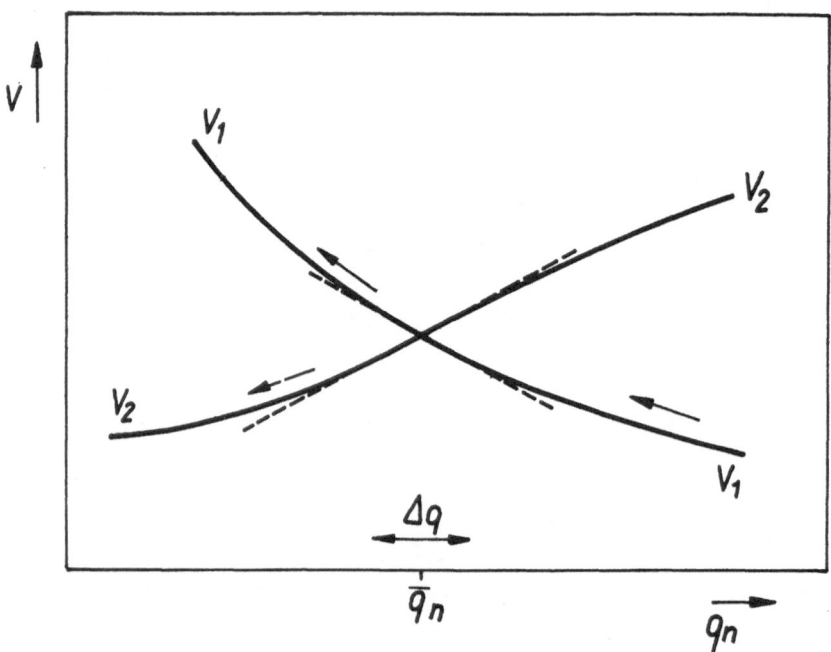

Fig. 5.-6

Near the intersection line the adiabatic potential surfaces V_1 and V_2 can be approximated by linear functions of a coordinate q_n normal to the line as shown in fig. 5.-6. We thus arrive at the model considered by Landau /41/:

$$V_1 = - F_1 \Delta q + \overline{V} ,$$
$$V_2 = - F_2 \Delta q + \overline{V} , \quad (\Delta q \equiv q_n - \overline{q}_n)$$
$$C_{12} = \overline{C}_{12} , \tag{2}$$

where $\bar{q}_n$ is the coordinate of the point of intersection, and $-F_1$, $-F_2$ are the derivatives of the electronic energies V_1, V_2 taken at that point (see fig. 5.-6). Finally, the equation of a trajectory passing close to the intersection point, in the vicinity of that point is approximated by a linear function of time:

$$\Delta q = u_n t \tag{3}$$

where u_n is the (constant) velocity describing the motion along the coordinate q_n.

First-order perturbation treatment when applied to eq. (5.3.-1) gives for the probability P_{12} of a non-adiabatic transition

$$P_{12} = \left| \int_{-\infty}^{\infty} \frac{\bar{C}_{12}}{i\hbar} \exp\left\{ \frac{i}{\hbar} \int_0^t (V_1(t) - V_2(t)) dt \right\} \right|^2 = \frac{2\pi |\bar{C}_{12}|^2}{\hbar u_n \Delta F_n} \tag{4}$$

where u_n and $\Delta F_n = |F_1 - F_2|$ depend on the position on the intersection line, whereas $\bar{C}_{12}$ depends on the position on the line as well as on dynamical parameters.

For a collision of two atoms (compare section 3.1.2.) we have $q_n = R$, and u_n stands for the radial velocity u_R; u_R and ω depend on the impact parameter $\cdot\, b$, according to eqs. (3.1.-18a,b),

$$u_R = u \left[1 - (b/R_c)^2 - \bar{V}/E \right]^{1/2}, \quad \omega = u b / R_c^2, \tag{5}$$

$E = \mu u^2 / 2$ being the initial relative kinetic energy. The expression (4) for the transition probability is valid provided $P_{12} \ll 1$; this condition is always satisfied if the linear approximation (2) holds in the range where the interaction is appreciable. In the model described by eqs. (2)

the two-atom system therefore tends to stay in its original adiabatic state with a probability $P_{11} = 1 - P_{12} \simeq 1$.

In fig. 5.-6, the motion before transition is represented by a full arrow on the right of the coupling region (extension Δq) whereas full and dashed arrows on the left of the coupling region represent adiabatic and non-adiabatic motions, respectively, after transition.

Upon substituting expressions (5) into formula (4), the transition probability can be expressed in terms of the impact parameter; the average cross section and the rate constant for the non-adiabatic process then can be calculated. It is important to realize that the intersection point must be crossed twice in atomic collisions: once when the atoms approach each other, and then when they fly apart. Thus in evaluating the appropriate cross section, one must use a resulting transition probability, the so-called two-way transition probability P_{12} (already mentioned in the preceding section) which is related to the one-way transition probability P_{12} by[1]

$$ \mathcal{P}_{12} = 2\, P_{12} \left(1 - P_{12}\right). \tag{6} $$

In a sense, this is similar to the classical approximation of the classical S-matrix method. To account for the interference, i.e. to make an approximation on the level of the primitive classical S-matrix approach, one should sum not probabilities but amplitudes; then one gets, instead of eq. (6),

[1] This expression results from the addition of the fluxes which correspond to the two different possible trajectories, one hopping at the first passage, the other at the second passage.

$$\mathcal{P}_{12} = 4\, P_{12}\, (1 - P_{12})\, \sin^2 \Delta\varphi \tag{7}$$

where the phase $\Delta\varphi$ is related to the difference of actions calculated along the two trajectories from the turning points to the crossing point. If $\Delta\varphi$ is large, averaging gives $\overline{\sin^2 \Delta\varphi} = 1/2$, and eq. (7) reduces to eq. (6). Since $P_{12} \ll 1$ in the case under consideration, $\mathcal{P}_{12} \approx 2 P_{12}$ is a good approximation.

Using expression (4) together with $\overline{C}_{12}$ from eq. (1) and taking into account the relations (5), we can calculate the corresponding cross section:

$$\sigma_{12}(u) = 2\pi \int_0^{R_c} \mathcal{P}_{12}(u,b)\, b\, db = \frac{16\sqrt{2}}{3}\, \frac{\pi^2}{\sqrt{\mu}\,\Delta F}\, \langle 1|\hat{\jmath}_\omega|2\rangle^2 \frac{(E-\overline{V})^{3/2}}{E}. \tag{8}$$

Furthermore, the rate constant for the non-adiabatic process can be obtained as

$$k_{12} = \left(\frac{8kT}{\pi\mu}\right)^{1/2} \pi R_c^2 \langle \mathcal{P}_{12}\rangle \exp(-\overline{V}/kT) \tag{9}$$

with the averaged transition probability

$$\langle \mathcal{P}_{12}\rangle = \frac{2\pi^{3/2}(2kT/\mu)^{1/2}\,\overline{C}_{12}^2}{\hbar\, R_c^2\, \Delta F}. \tag{9a}$$

The first factor in eq. (9) is the number of collisions corresponding to a cross section of radius R_c, calculated according to kinetic theory, and the last one is a typical Arrhenius factor. This latter appears because the two atoms must come to a distance R_c close to each other, overcoming in the process an energy of repulsion $\overline{V}$, for the non-adiabatic transition to occur; the quantity $\overline{V}$ therefore plays the role of an energy barrier.

5.4.2. Non-Adiabatic Transitions between Electronic Terms of the Same Symmetry

Though adiabatic terms of the same symmetry may cross (see section 5.1.), typically they will **pseudo-cross**. The $(s-1)$-dimensional line of pseudo-crossing (sometimes called the crossing seam) defines the region of possible non-adiabatic coupling. In the direction perpendicular to the seam, defined again as a coordinate q_n , profiles of the two adiabatic surfaces will look like hyperbolae, see fig. 5.-7. This pattern of the adiabatic terms V_1 and V_2 can be, and actually very

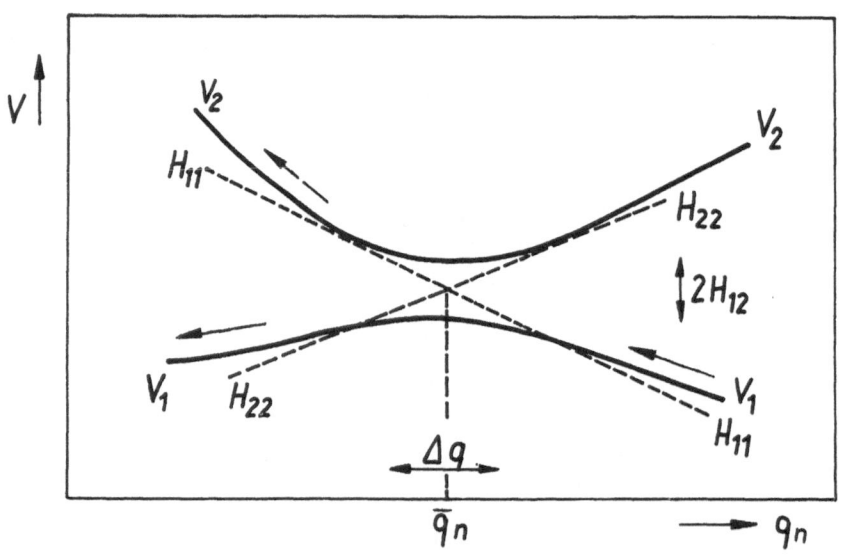

Fig. 5.-7

often is, related to a more simple behaviour of crude adiabatic terms V_1^o and V_2^o which cross.

Such crude adiabatic states can usually be obtained in a
natural way when potential-energy surfaces are derived in an
approximation which neglects some small part of the total
interaction. An example is provided by spin-orbit coupling;
it is indeed small compared to the Coulomb interactions of
nuclei and electrons for light atoms and even for atoms of
intermediate mass. Conservation of electron spin is satisfied on potential-energy surfaces obtained without taking
into account spin-orbit coupling; such surfaces can cross
each other but the crossing becomes a pseudo-crossing when
spin-orbit interaction is considered too.

In the Landau-Zener model of non-adiabatic coupling between terms of the same symmetry, the crude adiabatic terms
V_1^o and V_2^o are identified with diabatic terms V_1^d and
V_2^d, that is the dynamic coupling between the states V_1^o
and V_2^o is supposed to be negligible compared to the static
one. The model itself is formulated in terms of a diabatic
Hamiltonian defined by its matrix elements:

$$V_1^d = - F_1 \Delta q + \overline{V},$$

$$V_2^d = - F_2 \Delta q + \overline{V}, \qquad (\Delta q \equiv q_n - \overline{q}_n)$$

$$V_{12}^d = a = const. \tag{10}$$

For such a Hamiltonian the adiabatic terms and the dynamic-
coupling matrix element are determined (by diagonalization)
as

$$V_1 = \bar{V} - \tfrac{1}{2}(F_1 + F_2)\Delta q + \Delta V,$$

$$V_2 = \bar{V} - \tfrac{1}{2}(F_1 + F_2)\Delta q - \Delta V,$$

$$C_{12} = i \frac{a \Delta F \dot{q}_n}{(\Delta F \Delta q)^2 + 4 a^2}, \qquad (11)$$

with

$$\Delta V \equiv \left[(\Delta F \Delta q)^2 + 4 a^2 \right]^{\tfrac{1}{2}}, \quad \Delta F \equiv |F_1 - F_2|. \quad (11a)$$

The exact solution of the two coupled equations (5.3.–1) or (5.3.–9) in either representation with $\Delta q = \Delta q(t)$ defined by eq. (3) gives /35a/

$$P_{12} = \exp\left(-2\pi a^2 / \hbar \Delta F u_n\right). \qquad (12)$$

The ratio $2\pi a^2 / \hbar \Delta F u_n$ can be interpreted as the Massey parameter realizing that the characteristic length of the dynamic coupling as estimated from the expression for C_{12} given in eq. (11) is $\Delta q_c \sim 2a/\Delta F$, and the minimum separation of the adiabatic terms is $2a$:

$$\frac{2\pi a^2}{\hbar \Delta F u_n} \sim \frac{\Delta V_{min} \Delta q_c}{\hbar u_n}. \qquad (13)$$

Thus, in accordance with the general considerations (see section 2.2.2.) the transition probability is small for low velocities, i.e. if $u_n \ll (\hbar/\Delta V_{min} \Delta q_c)^{-1}$; for high velocities, $u_n \gg (\hbar/\Delta V_{min} \Delta q_c)$, the transition probability comes close to unity so that the system preferably follows the diabatic path.

Thus, for velocities high enough the diabatic terms can be interpreted as potential surfaces which govern the motion of the nuclei - in other words, the dynamic coupling between adiabatic states becomes so strong that it forces the system to follow the way on which it has its least possible value, i.e. the diabatic way. It is important to realize that away from the seam both potential surfaces, the diabatic and the adiabatic, will merge and be essentially the same. In fig. 5.-7 (like in fig. 5.-6), the motion before transition is represented by a full arrow on the right of the coupling region; full and dashed arrows on the left represent adiabatic and non-adiabatic motions, respectively, after transition.

For the transition probabilities between adiabatic levels and between diabatic levels, a simple connection can be established. If these transition probabilities are denoted by P_{12} and P_{12}^d, respectively, one has in fact

$$P_{12}^d = 1 - P_{12} . \tag{14}$$

If P_{12} is large, P_{12}^d must be small; expanding the exponential in eq. (12) one finds

$$P_{12}^d \approx (2\pi a^2 / \hbar \Delta F u_n) . \tag{15}$$

This expression coincides with eq. (4) if the parameter a is identified with the off-diagonal matrix element $\overline{C}_{12}$; such a result is as expected since in both cases one deals with transitions between intersecting levels.

Turning again to the case of atomic collision we calculate the two-way transition probability $\mathcal{P}_{12}$ according to eqs. (6) and (12) which gives

$$\mathcal{P}_{12} = 2 \exp(-2\pi a^2/\hbar \Delta F u_R)$$

$$\times \left[1 - \exp(-2\pi a^2/\hbar \Delta F u_R) \right]. \qquad (16)$$

In the literature this result is known as the Landau–Zener formula. A plot of $\mathcal{P}_{12}$ versus $x = \hbar \Delta F u_R/2\pi a^2$ is shown in fig. 5.-8 by the dashed curve with the maximum value $\mathcal{P}_{12} = 1/2$ corresponding to $P_{12} = P_{12}^d = 1/2$. The total cross section can be calculated in closed form for almost diabatic

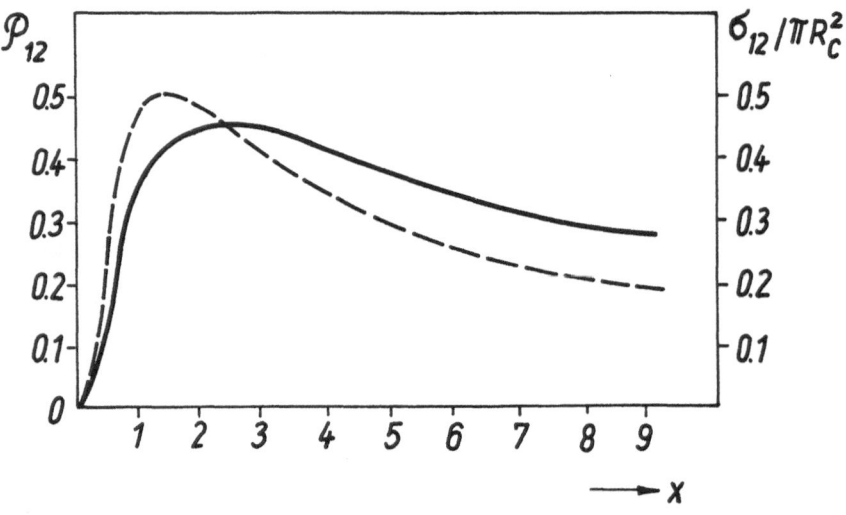

Fig. 5.-8

behaviour, i.e. for $P_{12}^d \ll 1$ when $P_{12} \approx 2 P_{12}^d$:

$$\sigma_{12}(u) = 2\pi \int_0^{R_c} 2 P_{12}^d(u,b) b\, db = \frac{4\sqrt{2\mu}\, \pi^2 a^2 R_c^2}{\hbar \Delta F} \frac{(E-\overline{V})^{\frac{1}{2}}}{E}; \quad (17)$$

it is drawn in fig. 5.-8 for $\mu u^2/2 \gg \overline{V}$ as a function of $x = \hbar \Delta F u / 2\pi a^2$ by the full line. The corresponding rate constant, if needed, can be calculated also /35a/.

5.4.3. The Surface-Hopping Trajectory Method

For a collision of two atoms, the probability P_{12} corresponding to a non-adiabatic transition from one adiabatic term to another one is simply related to the one-way transition probability by eqs. (6) or (7); during the collision the region of non-adiabatic coupling is traversed twice.

Going over to collisions between molecules, the theory of non-adiabatic transitions must be modified in two respects. First, the trajectory describing the relative motion of the nuclei in the range of significant non-adiabatic coupling can, in general, be arbitrarily oriented with respect to the line along which adiabatic surfaces cross or pseudo-cross each other. Second, even in the case of a single collision, the representative point in configuration space traverses the region of non-adiabatic coupling more than twice. Moreover, the trajectory has a different orientation with respect to the seam each time it approaches the transition region. As a result, the probability of a non-adiabatic transition P in a collision cannot be expressed in a simple way in terms of the analogous quantity P corresponding to a single passage through the region of non-adiabatic coupling. However, if P as defined by eq. (12) is used to calculate the one-way transition probability which results in traversing the seam, the dynamics of two colliding molecules can be described by the so-called surface-hopping trajectory approach.

Initially the representative point in configuration space moves on a particular potential-energy surface, along a certain trajectory which may pass through a region where the surface in question nearly crosses another one. If this happens, then the representative point can hop onto the latter surface with a certain probability P_{12} . Thus, when the representative point emerges from the region of non-adiabatic coupling, it can follow two trajectories, one on the initial potential-energy surface and another one on the neighbouring surface. These two trajectories move apart, and the system can be again described in terms of adiabatic motion on either potential-energy surface, until one of the trajectories brings the representative point once more into a non-adiabatic region, thus resulting in a new branching of trajectories. The successive occurrence of such cycles describing the non-adiabatic process leads to a redistribution of energy between electronic and nuclear degrees of freedom. Such an approach enables one to make maximum use of the theory of non-adiabatic transitions developed for atomic collisions and of the theory of inelastic and reactive collisions between molecules constructed within the framework of the adiabatic approximation (see chapter 3.).

The most serious objection to this kind of approximation is the inability of the method to describe interference effects and "classically forbidden non-adiabatic transitions". However, the former are believed to be unimportant because of averaging over initial vibrational and rotational phases, and the latter can be successfully dealt with in more sophisticated versions of the semiclassical theory /30/ by use of complex-valued trajectories which are "forced" to reach the seam and even to go smoothly from one adiabatic surface to another via their crossing point corresponding to complex-valued internuclear distances.

To illustrate the application of the surface-hopping trajectory approach let us briefly discuss the collision of a D^+ ion with a DH molecule /42/; there will be four channels for the process:

Table 5.-4: Semiclassical Treatment of Electronically
Non-Adiabatic Processes

Classical description of nuclear motion
on several potential energy surfaces
plus quantal treatment of the transitions
between them

Motion on a multi-valued
(branched) complex potential
energy surface,
along complex trajectories

Surface-hopping
trajectory approach

Only one out of many
trajectories used

Trajectory not accurately
determined
Interference neglected

Condition:

$$\Im 2\varphi/\hbar \gg 1$$
for this trajectory

$$\Delta V \ll E_{tr}$$

Probability small

Probability arbitrary

If the change of momentum, Δp, of the
nuclei in transitions from one potential
surface to another in the coupling region
is small compared to an average momentum p
and if the Massey parameter is large,
then

$$\Im 2\varphi/\hbar \approx l\Delta p/\hbar \approx l\Delta V/\hbar u \gg 1$$
and both approaches lead to the

Semiclassical adiabatic limit

$$P \sim exp(-l\Delta p/\hbar) \approx exp(-l\Delta V/\hbar u)$$

158

$$D^+ + DH \begin{cases} D^+ + DH & \text{channel a} & - & \text{elastic and inelastic scattering} \\ D\ + DH^+ & \text{channel a}^{\maltese} & - & \text{charge exchange without rearrangement} \\ H^+ + D_2 & \text{channel b} & - & \text{rearrangement (exchange reaction)} \\ H\ + D_2^+ & \text{channel b}^{\maltese} & - & \text{charge exchange with rearrangement.} \end{cases}$$

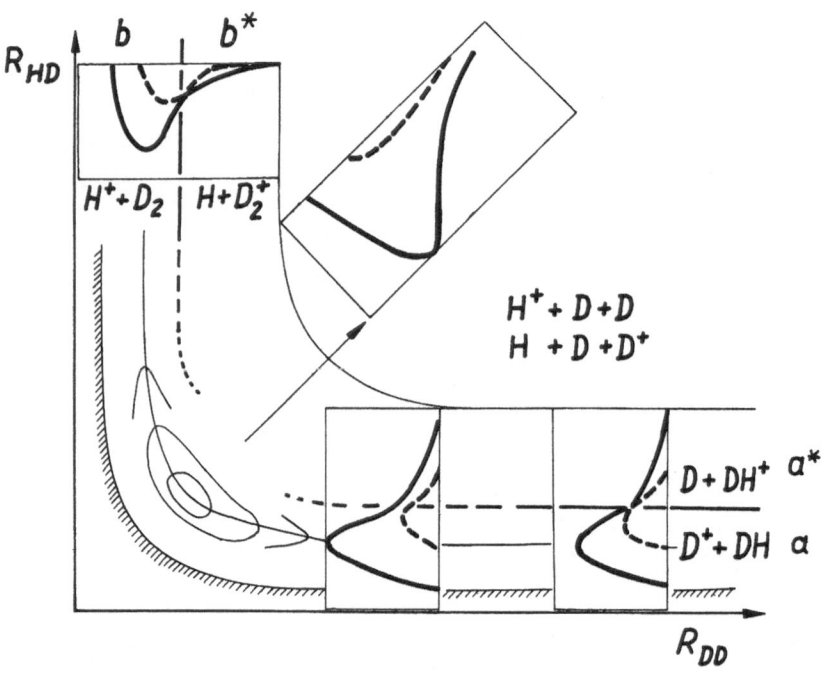

Fig. 5.-9

These channels are schematically shown in fig. 5.-9 displaying a qualitative view of the two lowest potential surfaces of the system DDH^+ for collinear nuclear configurations. The insertions represent profiles of the potential surfaces, the full line corresponding to the ground-state surface and the dashed line to the first excited one. The full lines in channels a, $a^{\ae}$ and b, $b^{\ae}$ and their (dashed and dotted) extrapolations to the region of small internuclear distances show the seam position with strong (full part), intermediate (dashed part) and weak (dotted part) non-adiabatic coupling. The strength of the dynamic coupling is related to the splitting of the adiabatic terms at the seam (as illustrated in the insertions) which varies from zero and very small values in the asymptotic regions (full part) via intermediate values (dashed part) to rather large ones at the region which corresponds to the stable complex DDH^+ on the lower surface.

Now, according to the theory, the motion of the system in the entrance channel at large R_{DD} is almost diabatic and corresponds to free vibrations either of DH or of DH^+. Potential curves reconstructed from the full and dashed parts (shown in the insertions) which join smoothly, just correspond to potential-energy curves of DH or DH^+ referred to the common asymptotic energy level of the dissociated system.

When the reagents D^+ and DH approach each other the static coupling between the two diabatic potential surfaces increases, and the probability P of the transition between the diabatic terms increases too. At sufficiently small R_{DD}, the probability will be of the order of unity, that is the initial diabatic behaviour changes into an almost adiabatic one; then the system DDH^+ moves independently either on the lower or on the upper surface until it is reflected back (channels a and $a^{\ae}$) or it reaches the exit valley. In the latter case, with R_{HD} increasing, the non-adiabatic coupling becomes appreciable, and for large R_{HD} this coupling is so strong that the system moves diabatically going out finally into the two uncoupled diabatic channels b and $b^{\ae}$.

This approach which proved to be successful in describing general features of such a simple non-adiabatic reaction /42, 43/ will be certainly very useful for learning more about the mechanisms of other non-adiabatic elementary chemical reactions (see, e.g. /44/).

Appendix

Transformation of the Hamiltonian to Center-of-Mass and Relative Coordinates

The separation of the center-of-mass coordinates in the Hamiltonian of a system consisting of atomic nuclei and electrons can be accomplished by introducing, instead of the laboratory coordinates ($\vec{R}_k$ for nuclei and $\vec{\xi}_\varkappa$ for electrons), the coordinates $\vec{S}$ of the center of mass, eq. (2.1.-1) and an arbitrary set of $(3N + 3N_e - 3)$ independent internal coordinates.

Using the internal coordinates /3a/

$$\vec{R}'_k = \vec{R}_k - \vec{R}_N \qquad (k = 1, 2, \ldots, N\text{-}1), \quad (1a)$$

$$\vec{\xi}'_\varkappa = \vec{\xi}_\varkappa - \vec{S} \qquad (\varkappa = 1, 2, \ldots, N_e), \quad (1b)$$

eq. (2.1.-2), and rewriting the Hamiltonian (2.-I), after simple straightforward calculations[1] one arrives at a kinetic-energy term of the form

$$T(R) + T(\xi) = T(S) + \left\{ T'(R') + T'(\xi') \right\} \quad (2)$$

---

[1] In the classical case, the Lagrangian function $L \equiv T - V$ has to be rewritten into the new (primed) coordinates and velocities leading, according to definition, to the new (primed) conjugate momenta.

In the quantum-mechanical case, the partial derivatives occurring in the Hamiltonian operator are to be converted, by means of the chain rule, to the new (primed) coordinates.

with (giving only the quantum-mechanical expressions)

$$\hat{T}(S) = -\frac{\hbar^2}{2M} \Delta_S \ , \tag{2a}$$

$$\hat{T}'(R') = -\frac{\hbar^2}{2} \sum_{k=1}^{N-1} \frac{1}{\mu'_k} \Delta'_k - \frac{\hbar^2}{m_N} \sum_{k<}^{N-2} \sum_{l=1}^{N-1} \vec{\nabla}'_k \cdot \vec{\nabla}'_l \ , \tag{2b}$$

$$\hat{T}'(\xi') = -\frac{\hbar^2}{2\mu'_e} \sum_{\varkappa=1}^{N_e} \Delta'_\varkappa + \frac{\hbar^2}{M} \sum_{\varkappa<}^{N_e-1} \sum_{\lambda=1}^{N_e} \vec{\nabla}'_\varkappa \cdot \vec{\nabla}'_\lambda \ , \tag{2c}$$

where $\mu'_k \equiv m_k m_N / (m_k + m_N)$ and $\mu'_e \equiv m_e M / (M - m_e)$ denote generalized reduced masses. The potential energy part, eq. (2.-Ic,d,e), does not contain the center-of-mass coordinates.

Another kind of internal coordinates has been proposed by FRÖMAN /3b/. He related the positions of all particles (nuclei and electrons) with exception of the N-th nucleus to the center of mass of the nuclei,

$$\vec{S}_n = \left(\sum_{k=1}^{N} m_k \vec{R}_k \right) / \left(\sum_{k=1}^{N} m_k \right) , \tag{3}$$

i.e. the internal coordinates are given by the relations

$$\vec{R}''_k = \vec{R}_k - \vec{S}_n \qquad (k = 1, 2, \ldots, N-1), \tag{4a}$$

$$\vec{\xi}''_\varkappa = \vec{\xi}_\varkappa - \vec{S}_n \qquad (\varkappa = 1, 2, \ldots, N_e), \tag{4b}$$

and $\vec{R}''_N = -\left(\sum_{k=1}^{N-1} m_k \vec{R}''_k \right) / m_N$. This transformation leads again to kinetic-energy terms of the form (2),

$$T(R) + T(\xi) = T(S) + \{T''(R'') + T''(\xi'')\} \quad (5)$$

with

$$\hat{T}(S) = -\frac{\hbar^2}{2M}\Delta_S , \qquad\qquad (5a)$$

$$\hat{T}''(R'') = -\frac{\hbar^2}{2}\sum_{k=1}^{N-1}\frac{1}{\mu_k''}\Delta_k'' + \frac{\hbar^2}{M_n}\sum_{k<l=1}^{N-2}\sum^{N-1}\vec{\nabla}_k''\vec{\nabla}_l'' , \qquad (5b)$$

$$\hat{T}''(\xi'') = -\frac{\hbar^2}{2\mu_e''}\sum_{\varkappa=1}^{N_e}\Delta_\varkappa'' - \frac{\hbar^2}{M_n}\sum_{\varkappa<\lambda=1}^{N_e-1}\sum^{N_e}\vec{\nabla}_\varkappa''\cdot\vec{\nabla}_\lambda'' , \qquad (5c)$$

where $\mu_k'' \equiv m_k M_n/(M_n - m_k)$ and $\mu_e'' \equiv m_e M_n/(M_n + m_e)$ denote generalized reduced masses, $M_n \equiv \sum_{k=1}^{N} m_k$ being the total mass of the nuclei. Again, the potential energy part does not depend on the center-of-mass coordinates.

The kinetic-energy terms (2a,b,c) and (5a,b,c) are of the same structure. The center-of-mass coordinates occur only in $\hat{T}(S)$ being thus separable; the remaining nuclear and electronic parts contain reduced masses and additionally so-called "mass-polarization terms" $\sim \vec{\nabla}_k\cdot\vec{\nabla}_l$ and $\sim \vec{\nabla}_\varkappa\cdot\vec{\nabla}_\lambda$.

Literature

/1/ ELIASON, M.A., and J.O. HIRSCHFELDER: J. Chem. Phys. 30 (1959) 1426.

/2/ BORN, M., and J.R. OPPENHEIMER, Ann. Physik (Leipzig) 84 (1927) 457.
BORN, M.: Nachr. Akad. Wiss. Göttingen, Math.-physikal. Klasse, Nr. 6 (1951).

/3/ a LUDWIG, G.: Die Grundlagen der Quantenmechanik, Springer-Verlag, Berlin-Göttingen-Heidelberg (1954), ch. XII.
b FRÖMAN, A.: J. Chem. Phys. 36 (1962) 1490.
c AROESTE, H.: Advan. Chem. Phys. 6 (1964) 1.

/4/ PAULY, H., and J.P. TOENNIES: Advan. Atomic and Molec. Physics 1 (1965) 195.

/5/ a PILAR, F.L.: Elementary Quantum Chemistry, McGraw-Hill, New York (1968).
b McWEENY, R., and B.T. SUTCLIFFE: Methods of Molecular Quantum Mechanics, Academic Press, New York (1969).
c ZÜLICKE, L.: Quantenchemie - Ein Lehrgang, Bd. 1, Grundlagen und allgemeine Methoden, VEB Deutscher Verlag der Wissenschaften, Berlin (1973).
d KUTZELNIGG, W.: Einführung in die Theoretische Chemie, Bd. 1, Quantenmechanische Grundlagen, Verlag Chemie, Weinheim (1975).

/6/ DEWAR, M.J.S.: The Molecular Orbital Theory of Organic Chemistry, McGraw-Hill, New York (1969).
POPLE, J.A., and D.L. BEVERIDGE: Approximate Molecular Orbital Theory, McGraw-Hill, New York (1970).

/7/ ELLISON, F.O.: J. Am. Chem. Soc. 85 (1963) 3540, 3544.
TULLY, J.C.: J. Chem. Phys. 58 (1973) 1396; 59 (1973) 5122.

/8/   KRAUSS, M.: Annu. Rev. Phys. Chem. 21 (1970) 39.
        LESTER, W.A., Jr. (Ed.): Proceedings of the
        Conference on Potential Energy Surfaces in Chemistry,
        IBM Res. Labor., San José, Calif. (1971).
        BALINT-KURTI, G.G.: Advan. Chem. Phys. 30 (1975) 137.

/9/ a MESSIAH, A.: Quantum Mechanics, Vol. I, II,
        North-Holland Publ. Comp., Amsterdam (1962).
     b DAVYDOV, A.S.: Kvantovaya Mekhanika, Nauka,
        Moscow (1973).
     c LANDAU, L.D., and E.M. LIFSHITZ: Kvantovaya Mekhanika,
        Fizmatgiz, Moscow (1963) [English transl.: Quantum
        Mechanics, Non-relativistic Theory, 2nd Ed., Addison-
        Wesley, Reading, Mass. (1965)].

/10/ a NIKITIN, E.E.: Ber. Bunsenges. phys. Chemie 72
        (1968) 949.
     b TOENNIES, J.P.: Ber. Bunsenges. phys. Chemie
        72 (1968) 927.

/11/   LANDAU, L.D., and E.M. LIFSHITZ: Mekhanika, Fizmatgiz,
        Moscow (1958) [English transl.: Mechanics, Addison-
        Wesley, Reading, Mass. (1960)].
        GOLDSTEIN, H.: Classical Mechanics, Addison-Wesley,
        Reading, Mass. (1950).

/12/   BERNSTEIN, R.B.: Advan. Chem. Phys. 10 (1966) 75.

/13/ a GOLDBERGER, M.L., and K.M. WATSON: Collision Theory,
        J. Wiley & Sons, New York (1964).
     b MOTT, N.F., and H.S.W. MASSEY: The Theory of Atomic
        Collisions, Clarendon Press, Oxford (1965).
     c NEWTON, R.G.: Scattering Theory of Waves and Particles,
        McGraw-Hill, New York (1966).
     d LEVINE, R.D.: Quantum Mechanics of Molecular Rate
        Processes, Clarendon Press, Oxford (1969).
     e NIKITIN, E.E.: Teoriya Elementarnikh Atomno-Molekul-
        yarnikh Protsessov v Gazakh, Khimiya, Moscow (1970).
        [English transl.: Theory of Elementary Atomic and Molec-
        ular Processes in Gases, Clarendon Press, Oxford (1974)].

/13/ f NIKITIN, E.E.: Theory of Elementary Atomic-Molecular
Reactions (Lecture Course), Parts I and II, Univ.
of Novosibirsk (1970) (in Russian).

g CHILD, M.S.: Molecular Collision Theory, Academic
Press, New York (1974).

/14/ a KARPLUS, M., R.N. PORTER and R.D. SHARMA: J. Chem.
Phys. 43 (1965) 3259.

b HEYDTMANN, G., in: Chemische Elementarprozesse (Ed.
H. Hartmann), Springer-Verlag, Berlin-Heidelberg-
New York (1968).

c BUNKER, D.L., in: Molecular Beams and Reaction
Kinetics (Ed. Ch. Schlier), Academic Press,
New York (1970).

d BUNKER, D.L.: Advan. Comput. Physics 10 (1971) 287.

/15/ SCHNEIDER, F., U. HAVEMANN and L. ZÜLICKE,
unpublished data.

/16/ GEDDES, J., H.F. KRAUSE and W.L. FITE: J. Chem.
Phys. 56 (1972) 3298.

/17/ BRUMER, P., and M. KARPLUS: J. Chem. Phys. 54
(1971) 4955.

/18/ a CHUPKA, W.A., J. BERKOWITZ and M.E. RUSSELL:
Abstracts of Papers, VI ICPEAC, Cambridge (1969).
D'AMICO, P.M.: Ion-Molecule Reactions, ACM Student
Report (1968).

b PACAK, V., K. BIRKINSHAW and Z. HERMAN: Abstracts
of Papers VIII ICPEAC, Beograd (1973).

/19/ a SCHNEIDER, F., U. HAVEMANN and L. ZÜLICKE: Z. phys.
Chemie (Leipzig) 256 (1975) 773.

b SCHNEIDER, F., U. HAVEMANN, L. ZÜLICKE, V. PACAK,
K. BIRKINSHAW and Z. HERMAN: Chem. Phys. Letters
37 (1976) 323.

/20/ a MUCKERMAN, J.T.: J. Chem. Phys. 54 (1971) 1155;
56 (1972) 2997.

167

/20/ b WILKINS, R.L.: J. Chem. Phys. 57 (1972) 912;
58 (1973) 3038; J. Phys. Chem. 77 (1973) 3081.

c BLAIS, N.C., and D.G. TRUHLAR: J. Chem. Phys.
58 (1973) 1090.
d JAFFE, R.L., and J.B. ANDERSON, J. Chem. Phys.
54 (1971) 2224.
e POLANYI, J.C., and K.B. WOODALL: J. Chem. Phys.
57 (1972) 1574.

/21/ a KOMPA, K.L., J.H. PARKER and G.C. PIMENTEL:
J. Chem. Phys. 49 (1968) 4257; 51 (1969) 91.
b POLANYI, J.C., and D.C. TARDY: J. Chem. Phys.
51 (1969) 5717.
c ANLAUF, K.G., et al.: J. Chem. Phys. 53 (1970) 4091.

/22/ SULLIVAN, J.H.: J. Chem. Phys. 46 (1967) 73.

/23/ RAFF, L.M., D.L. THOMSON, L.B. SIMS and R.N. PORTER:
J. Chem. Phys. 56 (1972) 5998.

/24/ a MAZUR, J., and R.J. RUBIN: J. Chem. Phys.
31 (1959) 1395.
b McCULLOUGH, E.A., and R.E. WYATT: J. Chem. Phys.
54 (1972) 3578, 3592.
c ZUHRT, Ch.: Ph. D. Thesis, Berlin (1974).
d ZUHRT, Ch., T. KAMAL and L. ZÜLICKE: Chem. Phys.
Letters 36 (1975) 396.

/25/ JOHNSON, B.R.: Chem. Phys. Letters 13 (1972) 172.

/26/ SCHATZ, G.C., J.M. BOWMAN and A. KUPPERMANN:
J. Chem. Phys. 58 (1973) 4023.

/27/ SAXON, R.P., and J.C. LIGHT: J. Chem. Phys.
56 (1972) 3874, 3885.
KUPPERMANN, A., G.C. SCHATZ and M. BAER:
J. Chem. Phys. 61 (1974) 4362.

/28/ a WOLKEN, G., and M. KARPLUS: J. Chem. Phys.
60 (1974) 350.

/28/ b KUPPERMANN, A., and G.C. SCHATZ: J. Chem. Phys.
        62 (1975) 2502; Phys. Rev. Letters 35 (1975) 1266.
     c ELKOWITZ, A.B., and R.E. WYATT: J. Chem. Phys.
        62 (1975) 2504.

/29/   KARPLUS, M., and K.T. TANG: Disc. Faraday Soc.
        44 (1967) 56.
        CHOI, B.H., and K.T. TANG: J. Chem. Phys. 61
        (1974) 5147; 62 (1975) 3642; 63 (1975) 2854.
        ZUHRT, Ch., F. SCHNEIDER and L. ZÜLICKE: Chem.
        Phys. Letters 43 (1976) 571.

/30/   MILLER, W.H.: Advan. Chem. Phys. 25 (1974) 69;
        30 (1975) 77.

/31/   FEYNMAN, R.P., and A.R. HIBBS: Quantum Mechanics
        and Path Integrals, McGraw-Hill, New York (1965).

/32/   HORNSTEIN, S.M., and W.H. MILLER: J. Chem. Phys.
        61 (1974) 745.

/33/   OVCHINNIKOVA, M.Ya.: Žurn. eksp. teor. fiz.
        67 (1974) 1276 (in Russian).

/34/   PECHUKAS, P.: Phys. Rev. 181 (1969) 166, 174.

/35/ a NIKITIN, E.E., in: Chemische Elementarprozesse
        (Ed. H. Hartmann), Springer-Verlag, Berlin-
        Heidelberg-New York (1968).
     b NIKITIN, E.E.: Advan. Quant. Chem. 5 (1970) 135.

/36/   BAZ', A.I., Ya.B. ZELDOVICH and A.M. PERELOMOV:
        Scattering, Reactions and Decays in Non-Relativistic
        Quantum Mechanics, Nauka, Moscow (1971) (in Russian).

/37/   CLARK, A.P., and A.S. DICKINSON: J. Phys. B
        4 (1971) L 112.

/38/   DICKINSON, A.S., and D. RICHARDS: J. Phys. B
        9 (1976) 515.

/39/   TAMIR, M.: Chem. Phys. Letters 28 (1974) 143.

        TOP, Z.H., and M. BAER: Chem. Phys. 10 (1975) 95.

        ZIMMERMAN, I.H., and T.F. GEORGE: J. Chem. Phys.
        61 (1974) 2468; Chem. Phys. 7 (1975) 323.

        BAER, M.: Chem. Phys. Letters 35 (1975) 112.

/40/   LONGUET-HIGGINS, H.C.: Proc. Roy. Soc. (London)
        A 344 (1975) 147.
        GEORGE, T.F., K. MOROKUMA and Y.-W. LIN:
        Chem. Phys. Letters 30 (1975) 54.

/41/   LANDAU, L.D.: Phys. Z. Sowjetunion 2 (1932) 46.
        ZENER, C.: Proc. Roy. Soc. (London) A 137 (1932) 696.
        STUECKELBERG, E.C.G.: Helv. Phys. Acta 5 (1932) 369.

/42/   PRESTON, R.K., and J.C. TULLY: J. Chem. Phys.
        54 (1971) 4297; 55 (1971) 562.
        KRENOS, J.R., R.K. PRESTON, R. WOLFGANG and J.C. TULLY:
        J. Chem. Phys. 60 (1974) 1634.

/43/   TULLY, J.C.: Ber. Bunsenges. phys. Chemie
        77 (1973) 557.

/44/   NIKITIN, E.E.: Uspekhi khimii 43 (1974) 1905
        (in Russian).

Recommended Supplementary Literature

Theoretical Foundation of Reaction Kinetics (Introductory Textbooks)

GLASSTONE, S., K.J. LAIDLER and H. EYRING: Theory of Rate Processes, McGraw-Hill, New York (1941).

LAIDLER, K.J.: Chemical Kinetics, McGraw-Hill, New York (1950).

LAIDLER, K.J.: Theories of Chemical Reaction Rates, McGraw-Hill, New York (1969).

JOHNSTON, H.S.: Gas-Phase Reaction Rate Theory, Ronald Press Comp., New York (1966).

KONDRAT'EV, V.N., and E.E. NIKITIN: Kinetics and Mechanism of Gas-Phase Reactions, Nauka , Moscow (1974) (in Russian).

Recent Advances in Methods and Applications

GEORGE, T.F., and J. ROSS: Quantum Dynamical Theory of Molecular Collisions, Annu. Rev. Phys. Chem. $\underline{24}$ (1973) 263.

PORTER, R.N.: Molecular Trajectory Calculations, Annu. Rev. Phys. Chem. $\underline{25}$ (1974) 317.

MICHA, D.A.: Recent Developments in the Theory of Reactive Molecular Collisions, Intern. J. Quant. Chem. $\underline{8}$ (1974) 263.

MICHA, D.A.: Quantum Theory of Reactive Molecular Collisions, Advan. Chem. Phys. $\underline{30}$ (1975) 7.

CLARK, A.P., A.S. DICKINSON and D. RICHARDS: The Correspondence Principle in Heavy-Particle Collisions, Advan. Chem. Phys. $\underline{36}$ (1977) 63.

TRUHLAR, D.G., and R.E. WYATT: H + H_2: Potential-Energy Surfaces and Elastic and Inelastic Scattering, Advan. Chem. Phys. $\underline{36}$ (1977) 141.

Subject Index

Action, classical, 102, 104, 105

Action-angle variables, 95, 103

Adiabatic,
  approximation, 7, 12, 14, 16–23, 123, 124, 125
  basis, crude, 127, 151
  electronic eigenfunctions, 126, 130
  potential-energy function 18, 126
  states, crude, 151

Airy function, 108

Angular momentum,
  electronic, 132
  nuclear, 35

Barrier,
  potential, 14, 23, 80
  rotational, 42
  square-well, 80

Born-Oppenheimer separation, 8, 16–23

Center of mass, 8
  coordinates of, 8, 161
  motion, 10, 19
  separation of, 8, 161–163

Centrifugal potential, 40

Charge transfer, non-resonant, 142

Classical,
  limit, 102, 103
  mechanics, validity, 33
  path approach, 109
  subsystem, 97, 109, 111

Collision,
  atom-atom, 34–44, 145, 133
  atom-diatomic, 44–58, 73, 89, 128, 134, 155
  collinear, 55, 73, 89, 113
  time, 51

Collisional complex, 14, 27, 28

Conservation of energy and angular momentum, 37

Coordinates,
  internal, 161, 162
  laboratory, 6, 161
  relative, 8, 161, 162
  relative, channel-adapted, 45

Correlation, electronic, 24

Correspondence principle, 97, 116

Coupling,
  localized, 124, 138
  region, 143

Cross section, 2, 42, 56
  differential, 4, 57, 84, 88
  differential elastic, 43
  reaction, 56, 88
  total, 4, 57
  Landau-Zener, 149, 155
Crossing,
  along a line, 128
  avoided, 127, 139-142,
      144, 150
  conic, 128
  covalent-ionic, 144
  pseudo-: see avoided
  seam, 150, 156

de Broglie wave-length,
  local, 33
Deflection,
  angle, 39, 42
  function, 40
Degree of freedom,
  external, 109
  internal, 109
Detailed equilibrium,
  principle of, 96
Diabatic basis functions,
      130, 137
Diatomics in molecules
  approach, 24
Distorted-wave approxima-
  tion, 93, 97
Dynamical operator, 71, 87

Ehrenfest equation, 69
Elementary process, 1
  complex, 28

direct, 28
dissociative, 2
elastic, 2
endoergic, 14
exoergic, 14, 62
inelastic, 2
non-adiabatic, 19, 123, 124
reactive, 2, 89
Energy, electronic, 12, 17
Equations of motion,
  canonical, 32, 37, 47
Exponential model, 142-143
External-force approximation, 113

Fixed-nuclei approximation, 17

Glory scattering, 44
Green's function, 84, 98
Green's operator, 85, 89

Hamiltonian, 6, 46, 47
  effective, 111
  electronic, 17
  symmetry of the, 125
Harmonic oscillator, forced,
      120, 121
Hartree-Fock approximation, 24
Heisenberg uncertainty re-
  lation, 65
Hydrogen exchange reaction,
      59, 74, 106
Hydrogen-iodine reaction, 62

Impact parameter, 38, 56
Initial state, 49
Interference effects, 105, 156

Internal,
  motion, 10
  states, 87

Lagrangian function, 35, 102
Landau-Zener formula, 154
Landau-Zener model, 139,
    145-155
  extended, 140-142
Laser, chemical, 62
Linear-exponential model,
    144-145
Lippmann-Schwinger equation,
    84, 85, 89

Mass,
  effective, 35
  reduced, 46
Mass-polarization terms, 162
Massey criterion, 135
Massey parameter, 22, 123,
    125, 152
Minimum-energy path, 14, 71
Molecular-beam experiment, 2
Monte-Carlo method, 50, 56

Non-adiabatic coupling, 129
  dynamic, 131, 133, 134
  matrix elements, 21
  operators, 18, 124
  rotational, 145
  static, 133
Non-adiabatic processes,
  rate constant, 149

Non-adiabatic transition,
    145-155
  classically forbidden, 142,
      156
  probability, 147, 152
Non-crossing rule, 127

Operator, time-evolution, 71,
    98
Orbiting, 42

Partial-wave expansion, 84
Partitioning of H, channel-
    dependent, 86
Path integral, classical, 110
Path-integral formulation
  of quantum mechanics, 103
Perturbation treatment,
  classical, 97, 100, 121
Potential curves, 12
Potential, effective, 111
Potential-energy function,
  adiabatic, 18, ·126
Potential-energy surface,
    11, 13, 123
  calculations, 23-26
  crossing, 125-128
Propagator, 98, 101

Quantum subsystem, 97, 109,
    111, 112
Quasiclassical approach, 49,
    95

Rainbow scattering, 44
Rate coefficient, 5, 149, 155
Reaction,
  bimolecular, 1
  elementary, 1, 5
  ion-molecule, 60
Reaction coordinate, 71
Reaction threshold, 90
Rebound mechanism, 31
Reflection,
  coefficient, 80
  probability, 83
Representation, diabatic, 137,
    151
Resonances, 34, 90, 96
Reversibility condition, 114

S matrix, 73, 117
  classical, 97, 99-109
Scattering,
  amplitude, 84, 88
  angle, 42
  channel, 26
  elastic, 34, 83-85
  glory, 44
  inelastic and reactive,
      44-63, 85-94, 106,
      155-160
  matrix: see S-matrix
  one-dimensional quantum,
      78-83
  operator, 73, 98
  rainbow, 44
  rearrangement, 44-58, 85-89
Scattering theory,
  classical, 34-57

  stationary quantum, 77-89
  time-dependent quantum,
      71-77
Schrödinger equation, elec-
    tronic, 17
Selection rules, 129, 133,
    134
Semiclassical approximation,
    20-23, 97, 109-115
Spin-orbit coupling, 126, 151
Stationary-phase method, 108
Stripping mechanism, 31, 60
Subsystem,
  classical, 97, 109, 111
  quantum, 97, 109, 111, 112
Superposition principle, 99,
    105
Surface-hopping trajectory
    approach, 155-160
Symmetry reduction, 127

Trajectory, 39, 51, 64, 95,
    97, 110, 111
  complex-valued, 106, 156
  initial conditions, 37, 47
Transition,
  amplitude, 111
  complex, 14, 27
Transition, non-adiabatic,
    21, 136, 145-155
Transition probability, 96,
    99, 113, 153
  non-adiabatic, 145-155
  one-way, 136, 148
  partially averaged, 118, 119
  two-way, 142, 148, 154

Transmission,
  coefficient, 80
  probability, 83
Tunneling, 34, 80, 83, 107,
    142
Turning point, 139, 142
Two-state model, 115,
    135-160

Uniform approximation, 108

Vibrational relaxation, 115

Wavepacket approach, 66-70, 71,
    73-77